POCKET

BOOK OF

KU-612-787

SCIENCE

ROBIN KERROD

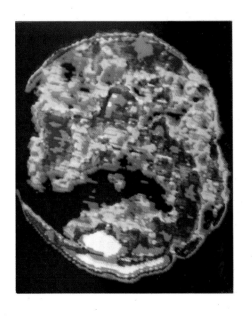

Kingfisher Books

Contents

POCKET

BOOK OF

SCIENCE

Kingfisher Books, Grisewood & Dempsey Ltd,
Elsley House, 24–30 Great Titchfield Street,
London W1P 7AD

First published in 1990 by Kingfisher Books
10 9 8 7 6 5 4 3 2
Copyright © Grisewood and Dempsey Ltd 1990

BRITISH LIBRARY CATALOGUING IN PUBLICATION DATA
Kerrod, Robin
   Science.
   1. Science
   I. Title   II. Series
   500
ISBN 0 86272 465 1

Edited by Mike Halson and Barbara Taylor
Picture research by Elaine Willis
Index by Isabelle Paton
Designed by David Jefferis
Phototypeset by Southern Positives and Negatives
   (SPAN), Lingfield, Surrey
Printed in Spain

Introduction

The *Pocket Book of Science* is an introduction to the world of science. In this fascinating world we can smash atoms apart; transmit pictures on radio waves and telephone calls on laser beams; look into the workings of the brain and manipulate the cells in our bodies; blast off into space and journey to the farthest reaches of the Solar System.

The word 'science' comes from the Latin word *scientia*, meaning knowledge. So in its broadest sense science encompasses all knowledge. However, we normally think of science as referring to that knowledge obtained by observation and experiment.

Scientists acquire and expand their knowledge in a systematic way, using what is called the scientific method. In this, they not only take note of *what* they see happen, they also try to find out *how* it happens and *why*. From their observations and experiments they can find the *what* and the *how*. They then use their powers of reasoning and past experience to develop a theory to explain the *why*.

To test their theory, scientists make a prediction about what should happen under given conditions, and this is called a hypothesis. They carry out further experiments to test their hypothesis. If the predicted and the actual results of the experiments agree, then their theory is correct. And they are able to

draw certain conclusions from it and maybe state certain scientific principles, or basic rules. If the results differ from those predicted, then their theory is wrong, and they must try to come up with another one.

Science advances our knowledge of the world about us, but does not directly bring us practical benefits. Only when that science can be applied and put to practical use do we benefit. And in general in the modern world, we usually come across science in its applied form, as technology. Technology, for example, takes the nuclear scientist's discovery

◄ **Painstaking laboratory** work occupies many chemists and biochemists for much of the time. They study and analyse chemical compounds, and determine their structure, using the latest instruments.

▲ **A marine biologist diving at** a coral reef. Field work – carrying out investigations of plants and animals in their natural setting – is particularly important in the biological sciences.

of the workings of the atom and produces nuclear power plants. It uses the unravelling of the mysteries of the substance DNA to produce highly advanced medicines.

The Birth of Modern Science

Historically, technology came before science. It really began when prehistoric people began shaping stones into tools. Prehistoric people had little time for pursuing scientific knowledge for its own sake – for observing, experimenting, thinking and

◄ **The chemical** sciences had their origin in the work carried out by the ancient alchemists. They strove for centuries to turn base (common) metals into gold and to find an elixir to prolong life. They didn't succeed, but they made many useful discoveries along the way.

► **Today the** computer is playing an increasingly important role in all the sciences. One exciting field is molecular modelling. Computer graphics are used to show the shape and structure of molecules.

such matters; theirs was a constant struggle for survival.

Only with the development of a more settled way of life in the early civilizations were people able to think about less practical matters. The first great thinkers were the ancient Greek philosophers. These people, who included Pythagoras, Plato, Aristotle and Archimedes, developed wide-ranging ideas on the nature of the world and of the universe.

Aristotle's contribution was particularly important, stressing the importance of practical observation as well as clever ideas. In the study of living things, he also established the principles of classification – arranging things into logical groups – something basic to biology even today.

The ancient Greek ideas dominated scientific thinking until the 1500s. The first person to present a challenge to them was Copernicus, who in 1543 proposed that it was the Sun, and not the Earth, that was at the centre of the universe. Astronomers such as Tycho Brahe, Johann Kepler and Galileo soon found plenty of evidence to support the new idea.

In the early 1600s Galileo became perhaps the first to apply the scientific method, establishing scientific principles such as the laws of the pendulum and falling bodies, as a result of careful observation, measurement and experiment. In the year Galileo died (1642) was born an

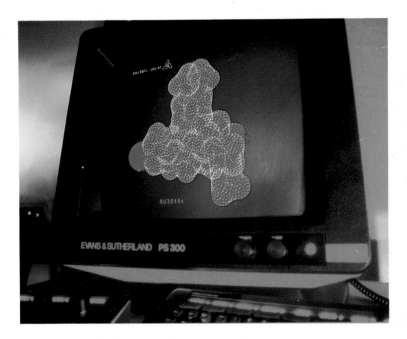

even greater scientist, Isaac Newton, who went on to develop the law of gravity, lay down basic laws of motion, pioneer the study of light, and invent a new kind of mathematics called calculus.

Newton published much of his work in a book, usually called the *Principia*. If we have to set a date for the beginning of modern science, the date of the *Principia*, 1687, would be appropriate.

Using this Book

The *Pocket Book of Science* covers the physical sciences, physics and chemistry, and also the science of life. Further volumes in the series deal with related sciences, such as astronomy and the study of animals.

The book is divided into nine main sections – *The Nature of Matter, Chemical Elements, Forces and Energy, Light and Radiation, Heat, Sound, Electricity and Magnetism, The Electronic World* and *The Science of Life.*

Each section is preceded by an introduction outlining the topics covered. Within the text, cross-references are given to other pages where further information on a topic may be found.

For general references, refer to the Index. Bold page numbers in the Index show where there is an entry in the Glossary, which provides a useful quick reference for topics that appear in the text.

The Nature of Matter

Every substance, every different material there is in the world around us is made up of stuff that we call matter. People, dogs, trees, rocks, oil, air: they are all made up of different kinds of matter. So are the Moon, the Sun, the stars, the galaxies and all the other heavenly bodies. These bodies float in a void where there is virtually no matter. We call this void space. The universe, which means everything that exists, is made up of matter and space.

We measure the amount of matter in a substance by its mass. For the same volume, or amount of space, some substances have more mass than others. We say they are more dense, or have a higher density.

But what is matter made of? If you could cut up a piece of matter into smaller and smaller and smaller pieces, you would eventually come to tiny particles that you couldn't cut up any more. We call these small specks of matter atoms. In most substances the atoms are linked together into larger particles, called molecules. It is the nature of the molecules, rather than the atoms, which gives a substance its properties.

Scientists once thought that the atom was like a ball which could not be divided. They now know that it is in fact made up of even smaller particles. And they have found out how to split it up. In doing so they have discovered ways of unleashing enormous energy. When this energy can be tamed, it is our friend. But when it is allowed to get out of control, it becomes an enemy that could threaten the future of all life on this planet.

▶ **A peaceful mountain scene** in the Tirol region of Austria. All the things you see are made up of different forms of matter. Some are living, others are non-living. Both living and non-living forms are made up of tiny atoms.

Solids, Liquids and Gases

In the world about us matter exists in one of three forms. It is either a solid, a liquid or a gas. These are called the three states of matter.

Rock is a typical solid: it is hard, and has a definite size and shape. Water is a typical liquid: it has a definite size but no definite shape. Air is a typical gas: it has neither a definite size nor a definite shape. It simply fills whatever space there is available. Liquids and gases are called fluids, because they flow readily from place to place.

Substances do not always remain in the same state. If you leave a saucer of water outside on a frosty night, the water freezes, or turns into a solid (ice). When the temperature rises, the ice melts into water again. If you heat up water in a kettle, it eventually boils. As it boils, the water gradually disappears. What happens is that the water slowly evaporates, or turns into a gas we call water vapour, which mixes with the air. When this water vapour is cooled by a cold window, for example, it condenses, or turns back again into liquid water.

So we see that water changes its state when the temperature changes. Most other substances will also change their state if the temperature changes enough. Solids will melt into liquids, and liquids will boil into gases as temperatures rise. On the other hand, gases will condense into liquids, and liquids will freeze into solids as temperatures fall. For each substance, melting, boiling, condensation and freezing occur at fixed temperatures.

Moving particles

What makes solids, liquids and gases so different? The answer lies in the different arrangement of the tiny particles of which they

◀ **Even solid rock will change** its state if the temperature rises high enough. This happens inside a volcano. When the volcano erupts, a fountain of molten rock spurts high into the air.

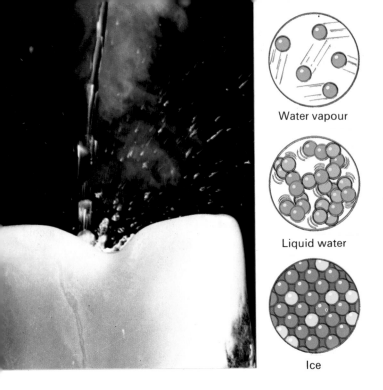

Water vapour

Liquid water

Ice

▲ We see one substance change its state often – water. In the picture, hot water in its liquid form is pouring on to ice, water in its solid form. Steam is also present where water vapour, water in its gaseous form, is condensing into little droplets.

The physical difference between the three states is explained by the way the water particles are arranged. In ice they are held rigid; in water they are less tightly packed together and can move around; and in water vapour they can move freely.

are made up. In a solid, for example, the particles are bound closely together by attractive forces (see page 26). They vibrate slightly, but cannot break loose from their neighbours.

When there is a rise in temperature, the particles gain more energy and vibrate more vigorously. Eventually, the particles

are able to move about by themselves. The solid melts into a liquid. As the temperature increases, the particles move faster. They break free from one another and begin escaping from the liquid to form a gas. This explanation of the behaviour of the different states is called the kinetic theory of matter.

Solids

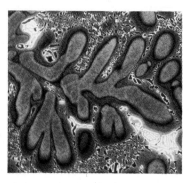

▲ **The crystals are clearly** visible in this sample of granite.

▲ **A magnified photograph of** tree-like metal crystals.

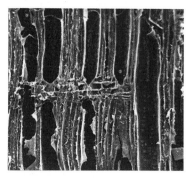

▲ **The fibrous structure of the** wood in the outer part of a tree.

Most of the substances we meet in our everyday lives are solids. Rocks and metals are familiar examples. A lump of rock has a definite size and a definite shape. It is hard, rigid and strong. It cannot easily be compressed (squashed), bent, stretched, twisted or otherwise forced out of shape. These properties make rock the perfect material for the foundations of massive structures such as skyscrapers and dams.

Metals, especially steel, are also valuable materials for building structures. They form the frames of skyscrapers and the towers of suspension bridges. Weight for weight, they are stronger than rock. In particular, they are better able to withstand tension, or pulling forces. Rock is strongest under compression, or when squashed. It is rather weak under tension.

Metal is also more elastic than rock. Most solids are elastic to some extent, which means they will deform slightly under pressure, but return to their original shape afterwards. This allows them to 'give' slightly under stress. Thanks to the elasticity of their metal frames, bridges and skyscrapers can safely sway in the wind. If they were totally rigid, stresses set up by winds would cause them to break up.

However, metals remain elastic only up to a certain point. If

► **A geode, or rock** cavity, filled with crystals of amethyst quartz. Note their regular shape. Amethyst is a semi-precious stone, widely used in jewellery.

they are stretched too much, they will be permanently deformed, and they may fracture. They may also fracture under much lighter loads if these loads are repeated time and time again. This sets up what is called metal fatigue.

Structures of solids

Most rocks contain many shiny coloured bits, which are crystals of the substances that make up rock. We call these substances minerals. In granite, for example, the main minerals are pink feldspar, black mica and white quartz. They are mixed haphazardly together and so the rock has no definite structure.

Also, granite has no definite composition. Two samples of the rock may contain feldspar, mica and quartz, but these minerals will be present in different proportions in each sample.

On the other hand, each mineral in a rock has both a definite structure and a definite composition. Quartz, for example, is a chemical compound (see page 50) that is always made up of one part silicon to two parts oxygen, no matter where it is found. As in all minerals, its atoms are arranged in a regular shape. or pattern, known as the crystal lattice.

In most rocks, the crystals are all squashed together and there is no hint of the lattice shape. But sometimes crystals are able to grow in holes in the rocks, without being squashed. Then they form regular shapes which are related to the crystal lattice structure. Quartz, for example, grows glassy, six-sided fingers.

Most other solids in fact are made up of crystals. All the metals are, for example, but the crystal shape is usually not obvious because they have been hammered, rolled, or poured in molten form into moulds. However, you can sometimes see feather-like crystals of tin on the inside of a can of fruit.

Solid materials made by living things, such as wood, do not form crystals. Wood has a regular structure, but it is not glassy. Instead it is made up of fibres.

Liquids

There is only one common liquid on Earth – water. And our everyday experience with water lets us build up a picture of the properties of liquids in general.

A liquid has a definite volume but not a definite shape. Pour water into a jar or a box, and it will take the shape of the container. The fact that you can pour water illustrates the fact that liquids can flow – they are fluids. The downward pull of the Earth's gravity (see page 72) makes them flow from a high to a low level. This is why rivers run downhill, and why there are waterfalls – the water is being pulled down by gravity.

Not all liquids flow as readily as water. Oil flows much more slowly and treacle more slowly still. They are said to have a greater viscosity than water, because their particles do not slide over one another as easily as water particles.

Tension at the surface

Fill a glass with water right to the brim, and then carefully add more water drop by drop. You will find that you can overfill the glass so that the water level rises above the rim. The water seems to be being held in by a kind of skin.

In fact, liquids do form a kind of skin at the surface. In the body of the liquid, the particles that make it up are surrounded on all sides by other particles. These all attract each other equally, so they are not pulled in any one

◀ **Rivers flow from a high** to a low level as a result of gravity, tumbling down cliffs as spectacular waterfalls.

▶ **Capillary action in a** narrow glass tube. When the tube is placed in a container of liquid, some of the liquid rises up it.

particular direction. At the surface, however, it is different. The particles receive a one-sided pull from those below because there are no particles pulling in the opposite direction. It is this one-sided pull that gives rise to the skin. We call this effect surface tension.

Defying gravity

Another effect comes into play at the surface of liquids in a container. Look at the surface of water in a tumbler, and you will notice that the surface curves up where it meets the glass. The reason this happens is that the particles of water are attracted

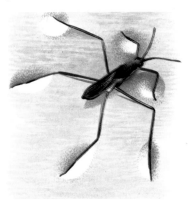

more strongly to the particles in the glass than they are to one another.

Mercury, another liquid, behaves differently. Its particles attract one another more than they attract glass. This means that when mercury is placed in a glass container, the surface of the mercury will curve down where it meets the glass sides. You can see this happening in a mercury thermometer (see page 110).

The attraction of particles of water (or another liquid) for glass and other substances gives rise to a further interesting effect known as capillarity. When water is contained in a narrow tube, the sides of the tube pull the water upwards.

Capillary action explains how towels dry our bodies. The fibres act like a mass of fine capillary tubes to soak up the water. In nature, capillary action takes place in plant stems. It draws water up from the ground through the roots to the leaves, so they can make their food.

◀ **A pond skater walking on** the invisible 'skin' on a pond, caused by surface tension.

Under pressure

The weight of the particles inside a liquid gives rise to an important property known as pressure. This exists anywhere below the surface, and is caused by the weight of the liquid particles pressing downwards. It is measured as the force acting on a given area – for example, 5 kilograms per square centimetre.

Pressure in a liquid increases with depth below the surface. Test this for yourself by filling a squeezy bottle with water and then making holes in it at intervals from top to bottom. Notice that the water spurts out farthest at the bottom – where the pressure is greatest.

Like solids, liquids cannot be compressed into a smaller space. When pressure is applied to a liquid, it is transmitted through-

▲ **Mechanical diggers at** work in roadbuilding. Their digging arms are operated by hydraulic rams, which move by liquid pressure.

out the liquid in all directions; this is why the water comes out sideways in the bottle experiment. This principle is used to deliver power in many machines. It is used to apply the brakes of cars, operate the landing gear of aircraft and power the arms of mechanical diggers. The field of study concerned with liquid pressure is known as hydraulics.

Floating and sinking

If you put a piece of wood in water, it floats. Push it down into the water and release it, and it bobs up again. The water obviously exerts an upward force on it, which we call upthrust. The upthrust just balances the weight of the wood.

If you put a brick in water, it sinks. But you notice that the brick feels lighter in the water than out. The upthrust makes the brick appear to lose weight. Naturally, the brick displaces, or pushes aside, some of the water when it is immersed. And if you weigh the water displaced, you will find that it equals the weight lost by the brick.

Archimedes' principle

The Greek scientist Archimedes made a general statement about sinking bodies in one of the best-known scientific principles. It states that when a body is immersed in a liquid, its apparent loss in weight is equal to the weight of the liquid displaced. When the body loses all its weight, it will not sink but float. Or we can say that a body will float when it can displace a weight of water equal to its own weight. This is the principle that enables massive steel ships to float – their shape enables them to push aside an amount of water weighing as much as they do. The size of naval ships is given in terms of their displacement – the weight of water they displace.

EUREKA! EUREKA!

Archimedes lived in Syracuse, Sicily, in the 200s BC, and was asked by the King of Syracuse to discover if his new crown was made of pure gold. One day, as he lowered himself into an overfull bath and watched the water slop over the sides, Archimedes realised that he could use the principle of displacement to solve the problem. Impure gold would displace more water than pure gold. He was so excited with his brainwave that he rushed naked into the streets, crying 'Eureka! Eureka!', meaning 'I have found it!'.

Gases

Gases are not as readily identifiable as solids and liquids. Nevertheless, they are just as real as the other two states of matter and likewise have their own typical properties.

Gases are invisible because their particles are far apart and shoot about by themselves. They do not cluster together as the particles do in solids and liquids. This means that gases are much less dense than solids and liquids. For example, 1 litre of iron weighs about 8 kilograms, or eight standard bags of sugar, while 1 litre of water weighs just one kilogram. But 1 litre of air weighs only 1.25 grams, which is about the weight of half a page of this book!

Gas particles travel at very high speeds. At room temperature, air particles, for example, travel at about 1670 kilometres per hour. Because of this, gases rapidly fill any container into which they are put. They have no size or shape of their own, unlike solids and liquids. They mix readily with any other gas present in the container, a process known as diffusion. Diffusion can take place over considerable distances. The scent of some female moths, for example, can diffuse through the air and attract males several kilometres away.

The most important gas

To us on Earth, air is the most important gas since without it we could not live. It surrounds the

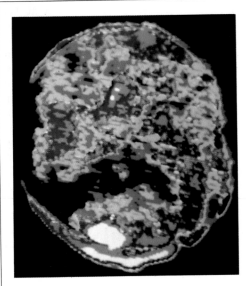

GAS BALL

This picture of the Sun was taken by astronauts on the Skylab space station. The Sun, like other stars, is a searing hot globe of gas. Inside, its temperature reaches tens of millions of degrees. At such temperatures ordinary atoms of gas cannot exist. Instead the atoms are split up, forming what is called a plasma. This is often called the fourth state of matter.

entire planet in a layer which we call the atmosphere. Most of the air is in the bottom part of the atmosphere, but traces are found up to about 500 kilometres high.

In fact, air is not really just one gas, but a mixture. It is made up mainly of nitrogen (78 per cent) and oxygen (21 per cent). Oxygen is the vital gas since it is this that we, and most living things, need in order to breathe.

The air also contains about 1 per cent argon, one of the noble gases (see page 49), as well as traces of water vapour and car-

◄ **A modern airship cruising** over London. Airships were once filled with hydrogen. It is the lightest gas but is highly flammable. Modern airships are filled with helium.

bon dioxide. The amount of water vapour varies, depending on the climate, while the level of carbon dioxide in the atmosphere is increasing as it is produced by the burning of many kinds of fuel. This is one of the causes of the 'greenhouse effect' (see page 83).

The two lightest and most common gases in the universe – hydrogen and helium – scarcely figure in air at all. Hydrogen is much lighter than air: 10 litres would not even weigh 1 gram. Helium is about four times heavier than this. Both gases have been used to provide the 'lift' in balloons and airships. Hydrogen is no longer used, however, because it catches fire readily. Helium does not.

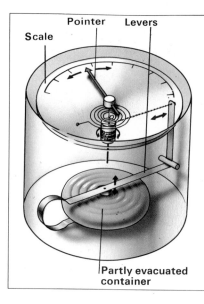

Scale Pointer Levers Partly evacuated container

BAROMETERS

We measure air pressure with an instrument called a barometer. A common type, called an aneroid barometer, is shown in the picture. It uses a container from which some of the air has been removed. The top of the container moves up and down as the air pressure decreases or increases. This movement moves a lever, which in turn moves a pointer over a scale.

A mercury barometer consists of a glass tube upside down in a reservoir of mercury. As air pressure increases, it pushes more mercury up the tube. A scale by the tube shows the current air pressure.

Applying pressure

A gas presses, or exerts pressure, on anything inside it, just as a liquid does (see page 18). The air above us in the atmosphere presses down with a force of 1 kilogram on every square centimetre of our bodies. We do not feel this force because the air presses evenly on all sides, inside and outside our bodies. The pressure gets lower as you climb higher into the atmosphere because there is less weight of air above.

Compressing gases

We can think of the pressure of a gas in a container as being caused by the gas particles colliding with the walls and rebounding back. On average at room temperature there are about 25,000 million million million particles present in every litre of gas.

Like liquids, gases can flow when they are pushed. But unlike liquids, gases can be compressed, or squeezed into a smaller area, because there is plenty of space between the particles. This enables us, for example, to pump air into bicycle tyres. By forcing in more air, you increase the number of particles inside the tyre. There are now more particles present to collide with the walls, and so the pressure increases – and the tyre becomes harder and harder.

Air resistance

For much of the time we are scarcely aware of the air around us. But when we are cycling into the wind, we feel the air resisting, or pushing against us, and we

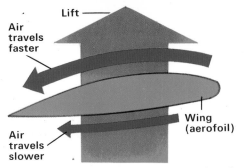

Lift

Air travels faster

▶ **Air travelling over** an aircraft wing, which has an aerofoil shape. Air travels faster over the top of the wing because of its greater curvature. This lowers the pressure there, and creates a lifting force.

Air travels slower

Wing (aerofoil)

have to pedal harder. Vehicles and aircraft, which travel much faster, use a lot of engine power in overcoming the resistance, or drag, of the air. To make them slip more easily through the air, they must be carefully shaped, or streamlined.

The study of the way air flows around things is known as aerodynamics. It is most important in the design of aircraft. Aircraft are designed with specially shaped wings, called aerofoils. When aerofoils travel through the air, more air is forced under the wing than over it. This leads to increased pressure below the wing, and decreased pressure above it. The resulting upward force is called lift. When an aircraft travels fast enough, the wings develop enough lift to support its weight. It then takes to the air and flies.

▼ **On racing cars, aerofoils** are fitted at the front and rear. They produce reversed lift, helping to keep the car firmly on the road.

It's a Small World

All substances, whether solids, liquids or gases, are made up of atoms. These are the smallest particles of a substance that can exist. Atoms are very tiny indeed. A single grain of sand contains billions upon billions of atoms.

For nearly a century scientists have been trying to work out what the atom is like. One thing they have proved is that 'atom' is not a good name for the smallest particles of matter! The word atom means 'that which cannot be divided', but under certain circumstances atoms *can* be divided. They are made up of smaller particles still.

We can picture the atom most simply as a miniature solar system. In our own Solar System, the Earth and eight other planets all travel through empty space around the massive body we call the Sun. Each planet circles the Sun at a different distance from it. Most of the Solar System is empty space. Similarly, most of the mass of the atom is located in the centre of the system, in the nucleus (plural nuclei). And moving around the nucleus in empty space are much smaller particles, called electrons.

The nucleus of an atom is made up of two main particles: protons and neutrons. Protons have a positive electric charge, while neutrons have no charge. Electrons have a negative electric charge. There are as many electrons outside the nucleus as there are protons inside it. The positive charge is thus equal to the negative charge, so overall the atom is electrically neutral.

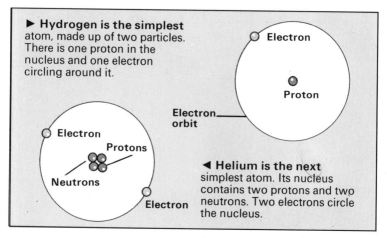

► **Hydrogen is the simplest** atom, made up of two particles. There is one proton in the nucleus and one electron circling around it.

Electron

Proton

Electron orbit

Electron

Protons

Neutrons

Electron

◄ **Helium is the next** simplest atom. Its nucleus contains two protons and two neutrons. Two electrons circle the nucleus.

THE STRUCTURE OF THE ATOM

A Greek named Democritos, in about 400 BC, was one of the first to think that matter was made up of atoms. However, this idea was not accepted until the early 1800s, after experiments carried out by an English chemist named John Dalton. But he, like Democritos, pictured atoms to be solid, like snooker balls.

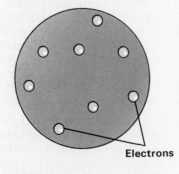

Electrons

◄ In this century scientists have come up with new pictures of the atom. The English physicist J. J. Thomson, who discovered the electron in 1897, pictured the atom as a kind of 'plum pudding', with electrons scattered in it like currants!

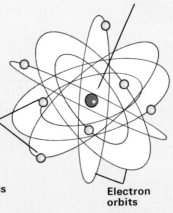

Nucleus

► In 1911, New Zealander Ernest Rutherford put forward the idea that the electrons circled around the nucleus in a cloud.

Electrons

Electron orbits

Electrons

Nucleus

◄ A more modern idea is that the electrons occupy a number of thin layers, or shells, around the nucleus – like the layers in an onion.

Electron shells

Inside the Nucleus

Just as the force of gravity holds the Solar System together (see page 72), so some kind of forces must hold the atom together. The electrons are kept in place inside the atom by electrical attraction, as they and the protons in the nucleus have opposite electrical charges, and opposite charges attract each other.

It is more puzzling what holds the nucleus together. This contains protons, all with the same positive electrical charge. And normally similar electrical charges repel (push away) one an-other. So why don't the protons fly apart? And what is it that holds the neutrons in place?

Scientists believe that the answer lies in a so-called 'strong force'. This is at least 100 times stronger than the electric force and thus overcomes the repulsion between the protons. It acts only within the nucleus.

The quest for quarks

Physicists have found evidence of the strong force in their experiments by means of atom-smashers. Using these powerful machines, properly called particle accelerators, they bombard atoms with high-speed atomic particles, such as electrons.

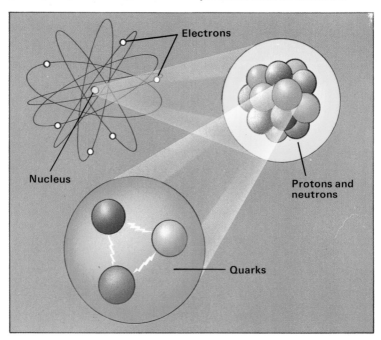

Electrons

Nucleus

Protons and neutrons

Quarks

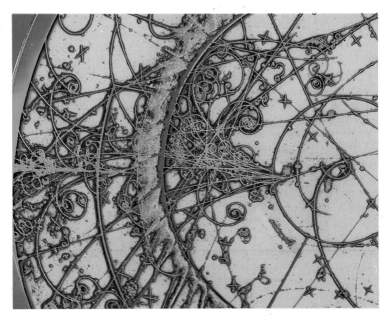

▲ **Tracks of sub-atomic**
particles, photographed in a
bubble chamber. The particles
make little bubbles appear in a
chamber of liquid hydrogen, held
near its boiling point.

They now realize just how complicated the atom is. Far from containing just three sub-atomic particles (that is, particles smaller than the atom), it contains many. In addition to protons, electrons and neutrons, more than 200 other particles have already been discovered, and each has a different mass (see page 74) and electric charge. Almost all of them exist for only a tiny fraction of a second.

Some are produced in pairs with their antiparticle, a kind of mirror image. An electron, for example, is often produced with its antiparticle, a positron. It is identical to an electron but has a positive, not a negative, charge.

As a result of their atom-smashing experiments, physicists have become convinced that most sub-atomic particles are made up of even smaller particles, called quarks. They now think that these are the basic particles of matter. However, most of the behaviour of ordinary matter can be explained in terms of the three main particles: protons, neutrons and electrons.

◄ **The atom is extremely**
complicated. The protons,
neutrons and electrons that make
it up seem, in turn, to be made up
of several kinds of basic particles
called quarks.

Unstable Atoms

Although there are millions of different substances in the world about us, there are only about 90 different kinds of atoms (see page 36). These combine together in varying ways to make millions of different substances – what chemists call 'compounds'. What makes one type of atom unlike another is the number of protons it has in its nucleus. The number of neutrons does not matter as far as the chemical properties of the atom are concerned. For example, chlorine atoms always have 17 protons. Commonly, they have 18 or 20 neutrons as well, but they are still chlorine atoms.

Most atoms remain the same all the time, no matter how they are combined in various substances. They are stable, or never changing. An atom of iron will always be an atom of iron, containing 26 protons. But a few atoms are unstable. They change as time goes by. A good example is the rare metal uranium. Over a period of time an atom of uranium will change into an atom of lead. This happens because uranium is radioactive: it gives off particles and radiation from its nucleus. The result is that the number of protons in its nucleus changes from 92 to 82. But a substance whose atoms have 82 protons in the nucleus is lead.

Alpha, beta and gamma

Ernest Rutherford, one of the pioneers of atomic research (see page 25), was also the first person to explain how radioactivity works, in 1902. For this he was presented with a Nobel Prize. He discovered that a radioactive

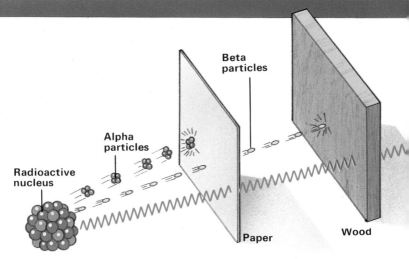

Radioactive nucleus

Alpha particles

Beta particles

Paper

Wood

substance gives off three kinds of radiation, which he termed alpha, beta and gamma.

Alpha radiation consists of a stream of particles made up of two protons and two neutrons. Beta radiation consists of a stream of electrons. Usually they are ordinary electrons, with a negative electric charge. But sometimes they are positrons, with a positive charge. Gamma radiation does not consist of particles. It is a form of electromagnetic radiation, rather like X-rays (see page 90).

All these forms of radiation can be dangerous to living things. In small doses they can cause sickness. In large doses they can kill. Carefully controlled, however, they can be used to kill tumours in cancer patients and for many other useful purposes.

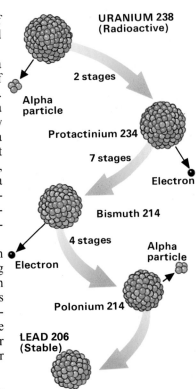

URANIUM 238
(Radioactive)

2 stages

Alpha particle

Protactinium 234

7 stages

Electron

Bismuth 214

4 stages

Alpha particle

Electron

Polonium 214

LEAD 206
(Stable)

▲ A simplified diagram showing how radioactive atoms break down, give off atomic particles and change into other atoms. Such a sequence is known as a radioactive series. The complete radioactive series in which uranium−238 breaks down into lead−206 involves some 14 breakdown stages.

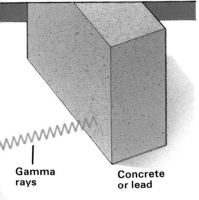

Gamma rays   Concrete or lead

◄ The three kinds of radiation given out by radioactive substances have different powers of penetration. A sheet of paper will stop alpha particles. And a plank of wood will stop beta particles. But a thick block of lead or concrete is necessary to stop gamma rays.

Energy from Atoms

As we have seen on page 29, uranium is an unstable atom. Its nucleus can break down at any time and give off radiation. But we can also force the uranium nucleus to break down artificially by bombarding it with neutrons. The nucleus of the atom takes in a neutron and then suddenly splits apart.

The interesting thing about this splitting process, called fission, is that enormous energy is given out. Scientists have found out how to tap this energy in the nucleus and put it to work. They put nuclear energy to good use in nuclear power stations (page 33). And they are also able to use it for destructive purposes in the terrifying atomic bomb.

▼ In nuclear fission, a neutron enters the uranium nucleus, which then splits apart. Energy and more neutrons are given off.

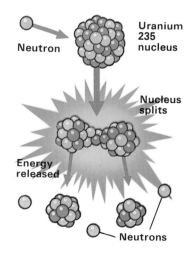

Neutron

Uranium 235 nucleus

Nucleus splits

Energy released

Neutrons

◄ **A mushroom cloud rises** out of the fireball created by the explosion of the first atomic bomb in New Mexico on 16 July 1945.

Chain Reactions

Bombarding uranium with neutrons from the outside only splits a few atoms to start with. But every time an atom splits, it gives off two or more neutrons of its own. If these then go on to split other atoms, which give off more neutrons to split still other atoms, a tremendous chain reaction will occur. And there will be a very rapid build-up of energy.

Such a build-up of energy occurs when the lump of uranium reaches a certain size, or critical mass. If the lump is smaller than this, too many neutrons escape, and a chain reaction does not take place.

In nuclear power stations, the chain reaction is always kept under control. In an atomic bomb, the chain reaction is allowed to go as fast as possible. American scientists exploded the first atomic bomb in New Mexico, USA, in July 1945. A month later one was dropped on the Japanese city of Hiroshima. It had the power of 20,000 tonnes of TNT, or ordinary high explosive, and flattened the city in an instant, killing over 100,000 people. This first alerted the world to the appalling horrors that a nuclear war would bring.

THE MIGHTIEST EQUATION EVER: $E = mc^2$

Just why does a uranium nucleus release so much energy when it splits? The reason is that a little mass is 'lost' in the process, and this 'lost' mass reappears as energy.

The first person to explain how this can be was the German-born scientist Albert Einstein (1879–1955). In his Special Theory of Relativity, published in 1905, Einstein showed that mass and energy are not separate things but are related.

Einstein worked out an equation that related energy (E) and mass (m). It was $E = mc^2$. c is the speed of light, which is over 300,000 kilometres (186,000 miles) per second. So c is a huge number, and c^2 is considerably greater still. In other words, the amount of energy released by the conversion of even a little mass is vast.

Nuclear fusion

The energy locked within the nucleus of an atom can also be released in another way. Some light atoms can be made to combine, or fuse together, to form heavier atoms. This process is called nuclear fusion. Energy is released when fusion occurs, because a little mass is 'lost' in the process, just as it is in nuclear fission.

We can think of nuclear fusion as the energy source of the universe, because it is the process that produces energy in the stars. In stars, atoms of hydrogen fuse to form helium. Here on Earth scientists have been able to imitate the process. But so far they can only do so in an uncontrolled and destructive way, in the form of the hydrogen bomb.

A hydrogen bomb uses two heavy forms of hydrogen for fusion, called deuterium and tritium. But these will only fuse at

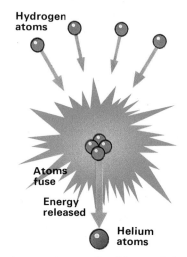

temperatures of millions of degrees. Only an atomic bomb can produce such temperatures, so one has to be used as a trigger for the hydrogen bomb.

Scientists have been trying for many years to control the fusion process so that they can use it to generate electric power. Some are using powerful magnetic machines called tokamaks to try to

◄ **The cluster of stars known** as the Pleiades, or Seven Sisters. Stars get their energy from nuclear fusion reactions going on in their interior at incredible temperatures.

► **In a reactor, nuclear fission** takes place in the fuel rods in the reactor core. It is kept under control by moving the control rods in or out. The moderator helps slow down neutrons so that they can more easily bring about fission. Heat is extracted from the core by a coolant, which gives up its heat in a heat exchanger, turning water into steam to drive turbines.

◄ **The principle of nuclear fusion.** Four hydrogen atoms are made to fuse, or combine together, to form a helium atom. Enormous energy is released in the process.

create the conditions necessary for fusion. Others are using lasers. But it will be a long time before fusion power plants become practical.

Nuclear reactors

Scientists in Chicago, USA, first achieved success in controlling nuclear fission in 1942. Under the direction of Enrico Fermi, they built a so-called atomic pile.

It was the first nuclear reactor, able to release nuclear energy in a controlled way.

Nuclear reactors are the key units in nuclear power stations. They allow the chain reaction in a critical mass of uranium to proceed at a controlled rate. The heat given out as the uranium atoms split is made to boil water into steam. And the steam is then used to spin the machines (turbogenerators) that generate electricity.

The chain reaction in the reactor is controlled by means of control rods, which reach down into the mass of uranium at the centre, or core, of the reactor. These rods are made of a

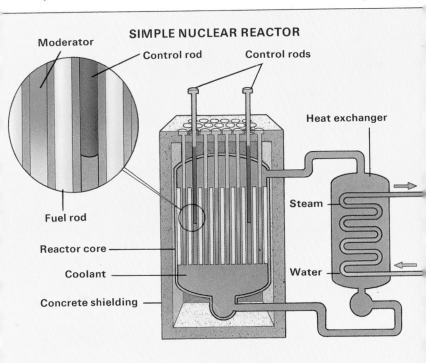

SIMPLE NUCLEAR REACTOR

Moderator

Control rod

Control rods

Heat exchanger

Steam

Fuel rod

Reactor core

Coolant

Concrete shielding

Water

material such as cadmium, which readily absorbs neutrons. When the rods are lowered fully into the core, they absorb so many neutrons that there are not enough available to bring about a chain reaction. To make the reactor start operating, the control rods are gradually raised. The number of neutrons increases until there are enough to bring about a chain reaction. And the reactor starts to produce heat. The position of the control rods is adjusted to give the required heat output. The reactor can be shut down at any time by pushing the rods fully home.

Heat is taken from the reactor core by means of a coolant. This

may be a liquid or a gas, which circulates through the core and carries the heat to a heat exchanger. There it gives up its heat to water, which turns to steam.

Some nuclear reactors contain a material called a moderator. Its job is to slow down the neutrons being produced. This is because slow neutrons split uranium atoms better than fast ones.

The neutrons escaping from the reactor core pose a hazard for living things. So do the other forms of radiation being given

THREE KINDS OF REACTOR

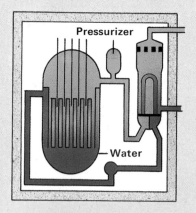

▲ The PWR (pressurized-water reactor) uses ordinary water as both a moderator and a coolant. The water circulates at very high pressure, which stops it turning into steam.

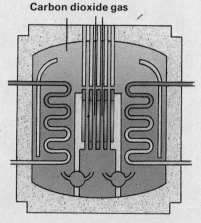

Carbon dioxide gas

▲ The AGR (advanced gas-cooled reactor) employs carbon dioxide gas as a coolant. It uses graphite, a form of carbon, as a moderator to slow down neutrons. It is found mainly in Britain.

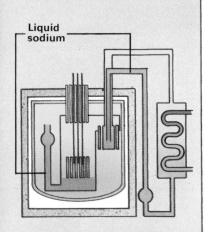

Liquid sodium

▲ **The breeder reactor** produces more fuel than it requires. It uses liquid sodium metal as a coolant. It needs no moderator because it uses fast neutrons to bring about fission.

out, particularly gamma radiation. That is why the reactor is enclosed in a thick biological shield, a massive container made of steel and concrete several metres thick. Another problem with nuclear reactors is that you have to get rid of the used fuel, and eventually of the reactor itself. As both are highly radioactive, and will remain so for thousands of years, this is difficult to do safely.

Of the three kinds of reactors mentioned in the box (left) the PWR is the most common. It is used in most of the world's nuclear power stations and also in nuclear submarines. Because reactors don't need air to work, nuclear submarines can stay underwater for months at a time.

Chemical Elements

If you look around you, you will see hundreds of different things, such as soil, leaves, plastic, rubber, glass, water, oil, wood and metal. Altogether in our world there are millions of different things made up of different materials. But all these millions of things are composed of only 92 simple substances. We call these the chemical elements. They are the basic building blocks of matter. The chemical elements are substances made up of the same kind of atoms. The way one atom differs from another is in the number of protons it has in the nucleus, and the number of electrons that circle around the nucleus.

Only a few elements exist in the pure state in nature. Gold is one: it is found in the ground. But most elements are found in substances in which they are combined with other elements. We call these substances chemical compounds. They are made up of two or more different elements joined together. The way the atoms in elements and compounds link together varies – it all depends on the arrangement of their electrons.

Of the 92 elements found naturally on Earth, most are solids, a few are gases, and just two are liquids – bromine and mercury. Most of the solid elements are metals. In addition to these 92, we now know of 17 more. Scientists have made these artificial elements by bombarding natural elements with atomic particles. All of them are radioactive, and some of them have a lifetime measured in fractions of a second.

▶ **Atoms of uranium in a** crystal of uranyl acetate, photographed by a scanning electron microscope. The colours are false; the magnification is nearly 100 million times. The other elements in the crystal, carbon and hydrogen, do not show up because they are transparent to the electrons used for imaging.

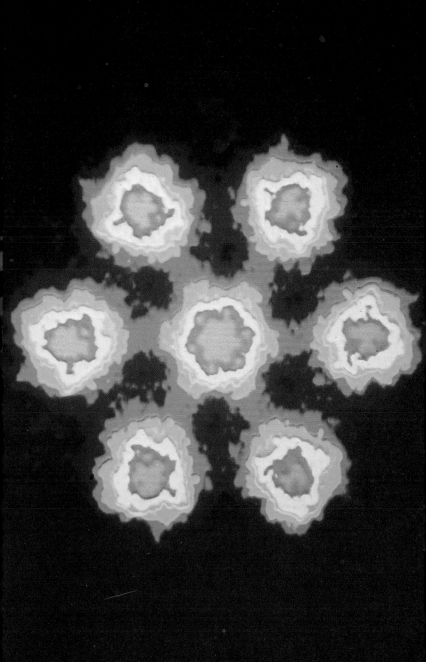

Grouping the Elements

Scientists began to realize that the chemical elements were the basic building blocks of matter in the 1800s. Chemists such as Sir Humphry Davy in Britain began discovering new elements and investigating the properties of others. They found that many could be grouped into families whose members had similar properties. And they began trying to find relationships between the elements in a family and between different families.

In their experimental work in the 1800s, chemists discovered how to calculate the relative weights of the atoms of the chemical elements. They called them atomic weights. These were expressed in relation to the weight of hydrogen. For example, they found that atoms of the element carbon are 12 times as heavy as atoms of hydrogen, so they gave carbon an atomic weight of 12.

The Periodic Table

A number of chemists began trying to put the elements in different classes according to their atomic weights and their properties. The person who was most successful in this was the Russian chemist Mendeleyev. He drew up a table in which he arranged the elements in order of their atomic weights in a series of horizontal rows and vertical columns. He arranged for elements of similar properties to be in the same

---

THE ELEMENTS AND THE PERIODIC TABLE

The Periodic Table is shown on pages 40 and 41. Each element appears in a box in the appropriate position in the table, which is determined by its atomic number. The element is identified by its chemical symbol.

For copper, the atomic number is 29; the symbol Cu.

Copper bracelet

column. Sometimes he had to leave gaps so he could position the elements sensibly. He then decided that these gaps must be the positions of elements not yet discovered.

Mendeleyev's arrangement is known as the Periodic Table. It still provides the basis for the modern method of classifying the elements. The modern Periodic Table, however, does not have the elements listed in order of their atomic weight (which we now call atomic mass). It has them listed in order of their atomic number. This is the number of protons in the atoms of the element. It is also the number of electrons which circle around the nucleus.

Periods and groups

One version of the Periodic Table is shown over the page. The horizontal rows are known as periods, and the vertical columns are groups. The modern group-numbering system used here is just one of a variety that have been used over the years.

If we ignore for the moment the extreme right-hand group (18), the most reactive of the elements (those that will combine or link up with other elements) are in the groups on either side of the table (1 and 17). Sodium and chlorine, for example, are both very reactive substances.

Also, the elements change in reactivity as you go down a group. For example, in group 1 potassium is more reactive than lithium higher up. On the other hand, in group 17 iodine is not as reactive as fluorine higher up. The physical properties of the elements also change going down a group. Densities increase, for example.

Of all the groups, the elements in group 18 have the most similar properties. They are all gases which react with other elements hardly at all. For this reason we call them the noble gases (see page 49).

The Periodic Table

▼ **In the Periodic Table,** hydrogen appears in group 1 and group 17 because it has similarities to both metals and non-metals (see page 48). The rare earths (lanthanides) are a collection of elements whose properties are so alike that they fit into a single block in the table. The actinides fit into a single block too for the same reason. They are radioactive. The elements after uranium (92) are man-made. Elements 105 to 109 have not yet been named.

1 Hydrogen H								
3 Lithium Li	4 Beryllium Be							
11 Sodium Na	12 Magnesium Mg							
19 Potassium K	20 Calcium Ca	21 Scandium Sc	22 Titanium Ti	23 Vanadium V	24 Chromium Cr	25 Manganese Mn	26 Iron Fe	27 Cobalt Co
37 Rubidium Rb	38 Strontium Sr	39 Yttrium Y	40 Xirconium Zr	41 Niobium Nb	42 Molybdenum Mo	43 Technetium Tc	44 Ruthenium Ru	45 Rhodium Rh
55 Caesium Cs	56 Barium Ba	57–71 Lanthanide series	72 Hafnium Hf	73 Tantalum Ta	74 Tungsten W	75 Rhenium Re	76 Osmium Os	77 Iridium Ir
87 Francium Fr	88 Radium Ra	89–103 Actinide series	104 Rutherfordium Rf	105	106	107	108	109
1	**2**	**3**	**4**	**5**	**6**	**7**	**8**	**9**

57 Lanthanum La	58 Cerium Ce	59 Praseodymium Pr	60 Neodymium Nd	61 Prometheum Pm	62 Samarium Sm	63 Europium Eu
89 Actinium Ac	90 Thorium Th	91 Protactinium Pa	92 Uranium U	93 Neptunium Np	94 Plutonium Pu	95 Americium Am

Legend

- Light metals
- Heavy metals
- Non-metals
- Rare earths (lanthanides)
- Actinides

					1 Hydrogen H	2 Helium He
		5 Boron B	6 Carbon C	7 Nitrogen N	8 Oxygen O	9 Fluorine F	10 Neon Ne
		13 Aluminium Al	14 Silicon Si	15 Phosphorus P	16 Sulphur S	17 Chlorine Cl	18 Argon Ar

28 Nickel Ni	29 Copper Cu	30 Zinc Zn	31 Gallium Ga	32 Germanium Ge	33 Arsenic As	34 Selenium Se	35 Bromine Br	36 Krypton Kr
46 Palladium Pd	47 Silver Ag	48 Cadmium Cd	49 Indium In	50 Tin Sn	51 Antimony Sb	52 Tellurium Te	53 Iodine I	54 Xenon Xe
78 Platinum Pt	79 Gold Au	80 Mercury Hg	81 Thallium Tl	82 Lead Pb	83 Bismuth Bi	84 Polonium Po	85 Astatine At	86 Radon Rh
10	**11**	**12**	**13**	**14**	**15**	**16**	**17**	**18**

64 Gadolinium Gd	65 Terbium Tb	66 Dysprosium Dy	67 Holmium Ho	68 Erbium Er	69 Thulium Tm	70 Ytterbium Yd	71 Lutetium Lu
96 Curium Cm	97 Berkelium Bk	98 Californium Cf	99 Einsteinium Es	100 Fermium Fm	101 Mendelevium Md	102 Nobelium No	103 Lawrencium Lr

The Light Metals

In the Periodic Table (pages 40–41), the elements are grouped into three broad classes: light metals, heavy metals and non-metals. The rare earths, or lanthanides, are metals with very similar properties. So are the actinides, which are all radioactive.

The light metals fall in the first two groups of the Periodic Table. In the first group (1), lithium (chemical symbol Li) has a few uses in metal form but is not common. Sodium (Na), one of the commonest and most reactive of the elements, also has few uses in metal form. One interesting application is as a liquid coolant in breeder reactors (see page 35).

Sodium is a very soft metal which you can cut with a knife. It is so light that it floats on water. It also attacks water vigorously. Bubbles of hydrogen gas are given off as the water (a compound of hydrogen and oxygen) is split up. Afterwards, the water left has a soapy feel. That is because it has been turned into an alkali (see page 58). The other

▼ **The famous white cliffs** along the English Channel are made up of chalk. Chalk is the commonest compound of the light metal calcium, which is much too reactive to be found by itself as a metal.

► **Firework rockets burst in** the night sky in a dazzling kaleidoscope of colours. The colours are due to compounds of light metals, such as strontium (crimson) and barium (green).

elements in this group also form alkalis in a similar way. That is why they are often called the alkali metals.

Sodium is by far the commonest of the alkali metals. Because it is so reactive, it is never found free in nature. We meet it in its most familiar form in salt, where it is chemically linked with chlorine (see page 50).

Among the most useful of sodium's other compounds is the alkali it forms with water, which is called sodium hydroxide. Its common name is caustic soda, because it is caustic, or able to burn flesh. Caustic soda is a very useful industrial chemical, used, for example, in making soap and manufacturing rayon.

The metals in the next group (2) also form alkalis. They are known as alkaline-earth metals. The alkalis they form are not as strong as those of the alkali metals. The most common alkaline-earth metals are calcium (Ca) and magnesium (Mg). They are found widely in compounds in the rocks. Calcium, for example, is found in chalk and limestone as the mineral calcium carbonate. Of this group, only beryllium (Be) and magnesium are used in metal form in lightweight alloys.

The Heavy Metals

This broad grouping of elements in the Periodic Table (pages 40–41) includes those we would call typical metals. Aluminium (Al) and iron (Fe) are examples. They are silvery white, shiny materials that are strong and hard. They are used to build all kinds of structures, from aircraft bodies to suspension bridges. For this reason they are both known as structural metals.

In their appearance, physical properties and use, the heavy metals are quite different from the light metals in the Periodic Table. But they share with the light metals two other properties that most non-metals do not have. They conduct, or pass on, heat and electricity very well (see page 53).

By far the best conductors of electricity are silver, copper and gold. These metals also have something else in common. They can be found in the form of metal in nature. For this reason they have been used for longer than any other metals. Because of their beauty and rarity, gold and silver have been prized down the ages for making jewellery.

Many very ancient gold and silver pieces have come down to us, thanks to another of their properties. They do not easily corrode. This means that they are not attacked by the oxygen or moisture in the air.

Other metals are readily attacked, for example iron. When iron is left to the mercy of the weather, it gradually corrodes. It combines with oxygen and moisture to form a red compound, which we call rust. In time the iron gets eaten away. Rusting can be stopped by coating the iron with paint or with another metal that does not corrode, such as tin or zinc. The coating of iron with a thin layer of zinc is called galvanizing.

Silver and gold are found in metal form in nature because they are not very reactive: they do not combine readily with other elements. A few other metals are like this, including platinum.

But iron and most other metals are more reactive. They are found in the ground only as chemical compounds in the form of minerals. The minerals have to be processed in some way to obtain the metal. Minerals from which metals can be extracted are known as ores. The most common ores are oxides, in which metals are combined with oxygen. They are generally processed by smelting – heating fiercely in a furnace with a substance such as coke (carbon). Iron is obtained in this way.

▲ **Steel girders form the** frame of all high-rise buildings. Steel is the most widely used of all metals.

▼ **Gleaming chromium covers** the bumpers and radiator of this classic Buick car.

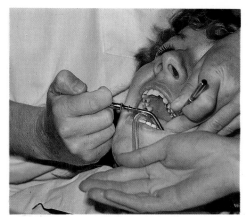

◀ **Metal alloy is** used to fill cavities in teeth. The alloy is made of an amalgam, or mixture with mercury, of silver or gold. When first mixed, the amalgam is pliable, but it soon sets hard.

Metal mixtures

It may seem strange, but in general pure metals are not as useful as impure ones. Many pure metals are quite soft and weak. But when other metals are mixed with them, they become very much harder and stronger. For example, copper and tin are both quite soft and weak metals. But when they are mixed together, they form a hard, tough material known as bronze. We call such a metal mixture an alloy.

Bronze is an alloy widely used, for example, to make coins and statues. It is of great historical importance because it was the first metal used on a large scale. This first happened in the Middle East in about 3500BC. It ushered in a period of history known as the Bronze Age. Copper forms

many other alloys, including cupronickel (with nickel), which is used to make so-called 'silver' coins. Brass is another common copper alloy (with zinc).

But the most useful alloy of all is made by adding small amounts of carbon – a non-metal – to iron. The result is steel. We might be tempted to say that steel provides the strength behind our whole civilization. That is how important it is to us. For example, most

▶ **Bronze has been used for** casting statues for thousands of years. It melts readily and flows easily into moulds. It also resists corrosion well.

▶ **Graphite is one of the two** natural forms of the non-metal carbon. (The other is diamond.) It is soft, and is mixed with clay to form the leads of pencils.

of our machines are made of steel, and we depend on them to produce the goods we need and to transport us about. No other metal is as cheap or plentiful, or has such a combination of useful properties. Its main drawback, rusting, can be overcome by alloying it with chromium and nickel, making stainless steel.

Non-Metals

In the Periodic Table (pages 40–41) the non-metals are found on the right-hand side. They include some of the commonest elements on Earth, such as silicon and oxygen. These two elements are often found combined together in nature in the form of minerals such as quartz (see page 15).

With nitrogen, which lies to the left of it in the table, oxygen makes up the air all around us. It is a very reactive element and will combine with most of the others, whether they are metals or non-metals.

To the left of nitrogen in the table is carbon, which is perhaps the most interesting of all the elements. This is because its

atoms can combine together in long chains and rings, something the other elements cannot do. In this way it can form millions of different compounds.

Many of these compounds are found in plants and animals. And it was once thought that they could only be made by living things. For this reason they were

▶ **The yellow non-metal** sulphur is found around vents from volcanoes.

called organic, and the study of them was termed organic chemistry. We still use the term today for the study of carbon compounds, even though we know a 'life force' is not always needed to make them.

As well as carbon, most organic compounds include hydrogen. Many include oxygen as well. Hydrogen is the lightest gas and has the simplest structure. It is difficult to know exactly where to place hydrogen in the Periodic Table, because in some chemical reactions it behaves like a metal, and in others like a non-metal. That is why it is shown in Group 1 (light metals) as well as in Group 17.

The halogens

The elements in Group 17 form one of the most closely knit families in the table. They are known as the halogens and have remarkably similar properties. For example, they are either poisonous gases (such as fluorine and chlorine) or they give off poisonous vapours (such as bromine and iodine). Bromine is one of only two elements in the table that are liquid. (The other is the metal mercury.)

Halogens form compounds with hydrogen, such as hydrogen fluoride and hydrogen chloride. When dissolved in water, these compounds become very powerful acids (see page 58), called hydrofluoric and hydrochloric acids. Hydrofluoric acid is the only substance that will attack

▼ To prevent the spread of diseases in a swimming pool, the water is treated with chlorine.

glass. The salts formed by the halogens are called halides. The silver halides are interesting because they are light-sensitive and form the basis of photography.

▲ **Mountaineers breathe pure oxygen at high altitudes because there is not enough present in the air around them.**

The noble gases

Group 18, on the right of the table, is also a very distinct family. Its members are similar because they combine with hardly any other elements and form hardly any compounds! For this reason they are called noble gases; they are also sometimes known as the inert or rare gases.

Traces of helium, neon, argon, krypton and xenon are present in the air, with argon being by far the most plentiful. Helium is also found in natural gas deposits. It is produced in the rocks from uranium. Uranium is radioactive and gives off alpha particles when it breaks down (see page 28). Alpha particles are the same as the nuclei of helium atoms.

The extreme inertness, or lack of activity, of the noble gases can be explained in terms of the structure of their atoms. They have electron shells (see page 25) that are full, with no extra electrons to lose or no room for extra electrons. So they cannot easily give or take electrons from other elements, as normally happens when elements combine.

Chemical Bonding

Two of the most reactive elements in nature are sodium and chlorine. (By 'reactive' we mean that these elements will combine, or link up with, many other elements, often with considerable violence.) In particular, they combine with each other to form the chemical substance sodium chloride, or common salt.

Salt is a chemical compound, the name we give to a substance formed when two or more elements combine together. In a compound the elements are not just mixed together. They are combined chemically and are linked by what are called chemical bonds. You cannot separate them as you can separate substances in a mixture – such as of marbles and chocolates.

The compounds that elements form when they combine have quite different properties from those of the elements themselves. Common salt, for example, could not be more different from the elements from which it was formed. Sodium is a soft metal and chlorine is a highly poisonous yellowish-green gas. Again, carbon, a solid, turns into a gas, the compound carbon dioxide, when it burns and combines with the oxygen in the air.

Full shells

How do elements actually combine with one another? As might be expected, combination occurs at the atomic level. And atoms combine together by using their electrons. They do not use all

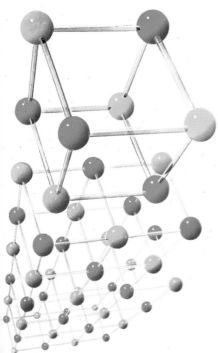

- Chloride ion
- Sodium ion

◄ **The elements sodium and chlorine** bond together to form the chemical compound sodium chloride, better known as common salt. The elements are present as ions (charged particles), forming an ionic compound.

their electrons, however – usually only those in the outermost electron shell (see page 25).

Each electron shell around the atom can hold only a certain number of electrons. When the shell is full, the structure of the atom is very stable. This explains why the noble gases are so stable – their electron shells are full.

So as a general rule we can say that when one element combines with another, it tries to end up with full electron shells. The number of electrons which an element uses when it combines is known as its combining power, or valency.

Ionic bonds

In many compounds, including common salt, the elements are linked by ionic bonds. Let us see what happens on the atomic scale. Sodium has a single electron in its outermost shell, while chlorine has seven. The easiest way for sodium to get full shells is

▲ **Salt pans on a Chinese** lakeside. The chemical compound we know as common table salt could hardly be more different from the elements, sodium and chlorine, which make it up.

for it to lose that one electron, because its inner shells are full. And the easiest way for chlorine to get full shells is to gain one electron, because the outer shell is full when it contains eight electrons.

The simplest way for both sodium and chlorine atoms to achieve full shells when they are brought together would be for the electron in the outer shell of sodium to transfer to the outer shell of chlorine. And that is exactly what happens.

Under normal conditions, the sodium and chlorine atoms are electrically neutral. They have as many protons with a positive charge as they have electrons with a negative charge. But when the atoms join, sodium loses an

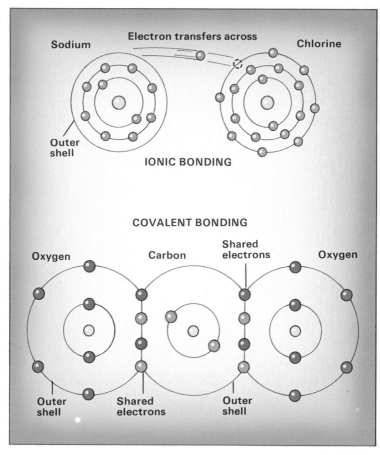

Sodium

Electron transfers across

Chlorine

Outer
shell

IONIC BONDING

COVALENT BONDING

Oxygen

Carbon

Shared
electrons

Oxygen

Outer
shell

Shared
electrons

Outer
shell

electron and thus a negative charge. So it ends up with an extra positive charge. We then call it a sodium ion.

Likewise, chlorine ends up with an extra electron and thus an extra negative charge. It becomes a chloride ion. Salt – the compound sodium chloride – therefore consists of a combination of positive sodium ions and negative chloride ions. Together

they form an ionic bond. The bonding is brought about by the electrical attraction of the positive and negative ions. All of the substances known as salts (sodium chloride is just one example) are ionic compounds.

Covalent bonds

Carbon and oxygen combine to form carbon dioxide in quite a different way. The carbon atom

◄ **Two common methods of** chemical bonding. In ionic bonding, electrons are transferred from one atom to another, and both become electrically charged ions. In covalent bonding, electrons are shared, with no change of charge.

has four electrons in its outer shell. To make a complete shell of eight, it has to share four electrons from other atoms. In carbon dioxide, it shares the electrons from two oxygen atoms, as shown opposite. At the same time each oxygen atom, which has six outer electrons, shares two of the carbon atom's electrons. So it too ends up with a full shell of eight. This method of bonding by sharing electrons is called covalent bonding.

Molecules

A group of atoms joined together by covalent bonds is called a molecule. The term is commonly used to refer to the smallest particles in a substance, but this is often incorrect. The particles may be atoms or ions instead.

Metallic bonding

Metals have their own kind of bonding, called metallic bonding. In a metal the atoms contribute their outer electrons into a common pool, or 'sea'. Since an electric current is a flow of electrons (see page 132), the presence of this sea explains why metals conduct electricity so well. This form of bonding also allows metals to be hammered and drawn into wire without breaking.

◄ **Using the latest** equipment, it is possible to spot the very atoms in some heavy metals. In this photograph of the metal iridium, the atoms show up as black dots. It was taken using an instrument called a field ion microscope.

Chains and Rings

Like the other elements, carbon can form simple compounds, such as carbon dioxide. The molecules, or basic units, of these compounds contain only a few atoms.

However, unlike the other elements, carbon can also form compounds whose molecules are made up of hundreds of thou- sands of atoms! The main feature of these compounds is that their molecules are made up of chains and sometimes rings of carbon atoms joined together. We call these compounds organic be- cause many are to be found in living things.

Among organic compounds there are many families whose members have similar properties. They differ from one another

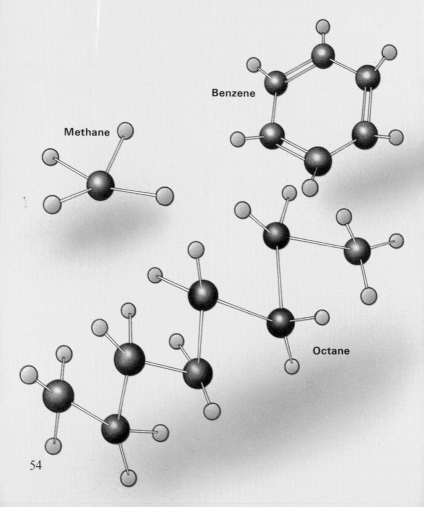

Benzene

Methane

Octane

only in the number of carbon atoms in the chain. The hydrocarbons are the simplest family. They are so called because they are made up of hydrogen and carbon atoms only. Methane (one carbon and four hydrogen atoms) is the first member of the hydrocarbon family. It is the main gas in the natural gas people use for cooking. Octane (eight carbon and 18 hydrogen atoms) is the eighth member. It is a liquid, found in crude oil or petroleum.

Petroleum is made up almost entirely of hydrocarbons. They also include benzene, a compound in which six carbon atoms are joined together in a ring. In general, compounds that contain a benzene ring structure have a powerful aroma, which is why we call them aromatics.

◀▼ A selection of different kinds of hydrocarbons. Methane, butane and octane are straight-chain compounds. Benzene is the best-known ring compound. Isobutane (or 2-methylpropane) is an isomer of butane, with the same number of atoms but a different structure.

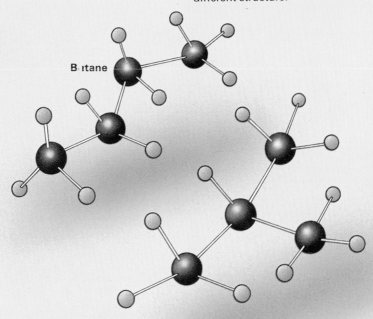

Butane

Isobutane (2-methylpropane)

Chemical Reactions

When you boil water in a saucepan, it gradually disappears. It turns into water vapour and escapes into the air. But if you hold a cold plate above the saucepan, the water vapour condenses on it – in other words it turns back into water. This conversion of water from liquid to vapour and back is an example of a physical change. The water does not change chemically.

When wood burns on a bonfire, it gives out heat and changes into ash. No matter what you do, you can't make it turn back into wood. This is an example of a chemical change. We say that the burnt up wood has undergone a chemical reaction.

What has happened is that the carbon and hydrogen that make up the wood have combined with the oxygen in the air to form new chemical compounds. They are carbon dioxide and water.

These reactions are examples of oxidation. This is a very common type of chemical reaction. The rusting of iron and steel is another example of oxidation. The orange-red rust that forms on the metal is a compound called iron oxide. It is the result of a chemical reaction between iron and the oxygen in the air.

Energy and Speed

In the above examples the chemical reactions of oxidation produce energy. We say they are exothermic – they give out heat. When wood burns, we can feel the heat given out. This is because the reaction proceeds quickly. We can't feel the heat from rusting, on the other hand, because the reaction takes place very slowly indeed.

Some reactions give out large amounts of energy at such a speed that they can lead to

◀ **Rusty cars in a scrapyard.** Rusting is a slow chemical reaction that takes place when iron is exposed to damp air.

▶ **Blowing up an old tower** block with explosives. An explosion is a chemical reaction that takes place very quickly. Large amounts of gases are produced, which expand rapidly and cause a shock wave.

explosions. Substances like dynamite and TNT (trinitrotoluene) release enormous energy and create explosions when they are made to decompose, or split up into simpler substances. Decomposition is another very common kind of chemical reaction.

Some chemicals can be made to react together more easily if another substance is added. For example, it is very difficult to make nitrogen and hydrogen by themselves combine to form ammonia. But if iron is present when the gases are brought together, they will combine much more easily. We call the iron a catalyst. It is a substance that changes the rate of a chemical reaction without being used up itself. Platinum is another very useful metal catalyst. In the body, enzymes act as catalysts.

CHEMICAL SHORTHAND

Chemists use a kind of 'shorthand' to show what takes place during a chemical reaction. They represent the atoms of the elements by their chemical symbol. H means one atom of hydrogen, C one atom of carbon, and O one atom of oxygen. H_2 means two atoms of hydrogen joined together – in other words, a hydrogen molecule. O_2 is two atoms of oxygen – an oxygen molecule – and so on.

Compounds are represented by a chemical formula, which shows the symbols of the atoms combined together and their relative numbers. The formula for the water molecule, for example, is H_2O; this shows there are two hydrogen atoms and one oxygen atom in every water molecule.

The chemical reaction that takes place when the carbon in wood combines with the oxygen of the air to form carbon dioxide can be represented by:

$$C + O_2 \rightarrow CO_2$$

This is an example of what is called a chemical equation. The oxygen is shown as O_2 because there are two atoms in the oxygen molecule, the smallest part of oxygen that can have a separate existence.

reacts with water. It is used to make soap. Caustic soda is in fact an alkali – a base that dissolves in water.

Neutralization can be used to relieve the pain of a bee sting (acidic) by applying bicarbonate of soda (basic). A wasp sting, on the other hand, is alkaline, and may be neutralized with lemon juice or vinegar.

When an acid combines with a base, it forms a compound called a salt, and also water. For example, hydrochloric acid (formula HCl) combines with caustic soda (NaOH) to form sodium chloride (NaCl). This is common salt (see pages 50–51) – the type used in cooking.

Hydrogen ions

Chemists have more precise definitions of acidic, basic and neutral substances. An acid is a substance that can give up hydrogen ions (H^+) – hydrogen atoms that have lost their single electron. A base is the opposite – it can accept hydrogen ions. This is how it neutralizes an acid.

All acids form ions when they dissolve in water. Hydrochloric acid, for example, forms hydrogen ions (H^+) and chloride ions (Cl^-). An alkali also contains ions. When sodium hydroxide dissolves in water, it produces

Acids, bases and salts

Another common kind of chemical reaction is that between an acid and a base. It is called neutralization. A substance that is neither acidic nor basic is said to be neutral.

Acids are very common chemical compounds: lemons, for example, contain citric acid, while vinegar contains ethanoic (acetic) acid. The sharp, sour taste of these substances is typical of acids.

Acids have many uses. Our stomachs contain hydrochloric acid, which helps digest food. Sulphuric acid and nitric acid are two common acids used to make fertilizers and other products.

Bases are chemical compounds that are the opposite of acids – they react with acids and neutralize them. A common base is caustic soda, or sodium hydroxide, formed when sodium

sodium ions (Na^+) and hydroxide ions (OH^-).

When an acid and alkali come together, the ions, as it were, swap over. The result is a salt, such as sodium chloride ($NaCl$), and water. We can represent what happens like this:

$Na^+ + Cl^- \rightarrow Na^+Cl$ (salt)
$H^+ + OH^- \rightarrow H_2O$ (water)

pH

In solution (dissolved in a liquid), hydrochloric, nitric and sulphuric acids are split up completely into ions. They produce a large concentration of hydrogen ions, which are available for neutralization. We call them strong acids. Citric acid, on the other hand, only produces a low concentration of ions, and we call it a weak acid.

Alkalis are classed in a similar way. Sodium hydroxide ionizes completely and produces a high concentration of hydroxide ions (OH^-). We call it a strong alkali.

The strengths of acids and alkalis are measured on the pH scale. It goes from 0, for a strong acid, to 14, for a strong alkali. A neutral solution – neither acid nor alkali – has a pH of 7.

ACID RAIN

In many parts of the world, forests and their wildlife are dying because of acid that rains down on them from the skies. The source of this acid is the burning of fossil fuels, which give off sulphur dioxide, nitric oxide and nitrogen dioxide. These gases turn into droplets of dilute acid in moist air, and can travel long distances before falling as rain or snow.

The Chemical Industry

Most of the materials we use have been processed in some way by chemicals. And a vast industry has grown up to produce chemicals.

The chemical industry is split up into many branches, which manufacture different products. Some branches are involved in producing basic chemicals; others in using these chemicals to make finished products. Major products of the chemical industry are sulphuric acid, fertilizers, plastics, synthetic fibres, pesticides, paints, detergents and drugs.

Sulphuric acid is the most widely used of all industrial chemicals, which is why it is often called the 'lifeblood of industry'. It is an example of a heavy

PETROLEUM REFINING

The raw material processed in a refinery is thick black crude oil, or petroleum. It is a mixture of thousands of hydrocarbon compounds which has accumulated in traps underground. Oil wells bring it up to the surface.

In the refinery, the hydrocarbon mixture is first turned into vapour by being heated in a furnace (1) to about 400°C.

The oil vapour passes into a distillation tower (2). This has about 40 trays across it, each kept at a different temperature – coldest at the top, hottest at the bottom. The vapour condenses into a number of hydrocarbon 'fractions', depending on the temperature of the trays.

Bubble caps (3) in the trays force the rising vapour to bubble through liquid that has already condensed. This helps to ensure that the fractions are efficiently removed from the vapour.

The boxes list some of the ways in which the fractions are put to use by the chemical industry.

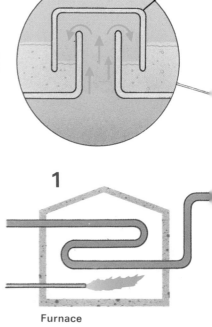

3 Bubble cap

1

Furnace

chemical – one produced in vast quantities. Ammonia and caustic soda are two other heavy chemicals. All three are examples of inorganic chemicals. They are derived from minerals or from gases.

In contrast, the chemicals used to make plastics, dyes and many other products are organic. The main raw material for the organic chemical industry is petroleum, or crude oil. The processing of petroleum into useful products takes place in an oil refinery. Coal used to be the main source of organic chemicals, and will be again when world supplies of petroleum run out sometime in the next century.

Distillation

The way a refinery operates and the processes it carries out are typical of the chemical industry as a whole. The first refinery process is fractional distillation. It is a process that separates the

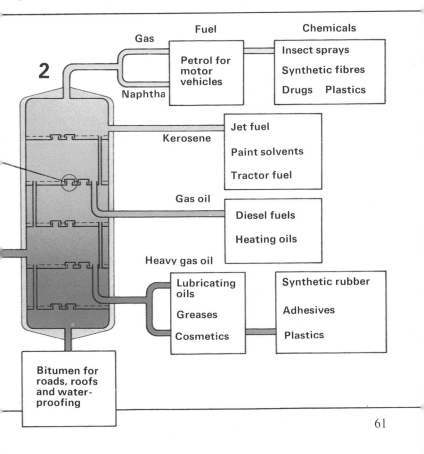

crude oil into the various hydro-carbons (see page 55) which make it up. They can be separated because they have different boiling points.

In the distillation process the oil is heated and turned into vapour. The vapour is fed to a tower, which has trays maintained at different temperatures. As the vapour passes through the tower, the hydrocarbons condense, or turn back into liquid. They do so at a level where the temperature is just below their boiling point.

In this way the oil is split up into a number of fractions. Many are used as fuels. Others go for chemical processing such as cracking, in order to turn them into useful chemicals.

▲ A plane sprays crops with fungicide to protect them against disease. Chemical sprays are among the most useful products made from petroleum chemicals.

Cracking

The least useful fractions are the heavier oils from near the bottom of the tower. But these can be converted into lighter, more useful fractions by cracking. This is a process that 'cracks', or splits, their large molecules into smaller ones. The new products may be used as fuels, or as chemicals for industry.

Some of the light gases that come off the top of the tower have molecules which are too small to be useful. So they are converted into substances with larger molecules. This process is called polymerization.

Polymerization is the main process used in the plastics industry. Is it used to build up simple molecules into a molecule with a very long chain of carbon atoms. It is the long chain of their molecules that gives plastics their properties. One main property of all plastics is that they can be easily moulded into shape when they are hot.

The best-known plastic is polyethylene (polyethene), often called polythene. It is made from ethene (ethylene), which is a light gas produced in oil refineries. It is a particular type of plastic called a thermoplastic, which softens when heated. PVC (polyvinyl chloride) and nylon are also thermoplastics. Nylon is especially useful because it can be readily pulled out into finer and finer fibres. It is one of the most important of all synthetic (artificially made) fibres.

The first plastic ever made, bakelite, is an example of an entirely different type of plastic. It is a thermosetting plastic – one which sets rigid when first moulded and cannot be remelted. It is named after Leo Baekeland, who first made it in the United States in 1909.

Bakelite is heatproof and is a good electrical insulator. It is made from phenol and formaldehyde. Melamine-formaldehyde is a similar modern material, and is widely used to manufacture heatproof kitchen worktops.

DETERGENTS

Detergents are synthetic products made from petroleum chemicals. Their molecules have a water-loving 'head' and a water-hating 'tail'.

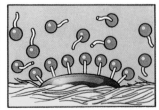

▲ **The water-hating, or** hydrophobic, tail is a long-chain hydrocarbon molecule. In water, it attaches itself to greasy dirt particles on fabrics.

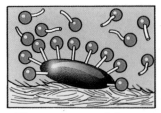

▲ **The greasy particles soon** become surrounded by detergent molecules, with the water-loving, or hydrophilic, heads on the outside.

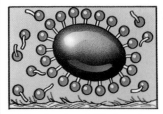

▲ **The hydrophilic heads** are attracted to the water molecules, and pull the greasy particles away from the fabric into the water.

Forces and Energy

Our lives, our world and indeed the whole universe are ruled by forces and energy. By forces, we mean pushes and pulls, kicks, twists and shoves. Almost always, forces cause something to move, or change the way they are moving. The science concerned with forces and energy, and the way they affect matter, is physics. Physics and chemistry investigate how matter is made up. Together they are called the physical sciences.

Imagine you are playing football on a windy day and kick the ball high in the air to another member of your team. Your kick is the force that gets the ball moving. The ball does not go on moving in a straight line. The force of the wind blows it off course. The force of resistance of the air slows it down. And the force of gravity – the Earth's pull – makes it fall to the ground. When it rolls along the ground, it slows down even more because of the friction, or rubbing, of the grass.

So even such an apparently simple action of kicking and passing a ball is influenced by all kinds of forces, which determine how the ball is going to move.

In order for the wind to blow, or your foot to kick, energy must be supplied. The energy that causes the winds to blow comes from the Sun in the form of heat. When you kick, you use up energy that comes from the food you eat. It is chemical energy produced when the sugars in food break down. There are lots more forms of energy in the world about us – for example, electrical energy (the most convenient kind), magnetic energy, and nuclear energy from the nuclei of atoms.

We can harness the various forms of energy and put them to work by using machines. Engines are machines that change energy directly into mechanical motion to drive other machines.

▶ **Space shuttle *Discovery*** lifts off with all engines blazing. The force of gases shooting backwards from the engines produces a force forwards that propels it into the sky.

Movement

Place a tennis ball on a table and leave it alone. What happens? It stays just where it is. Prod it with your finger and it will move. It will carry on moving straight, unless you prod it again, in which case it will move off in a slightly different direction.

In this simple scientific experiment you have shown that: (1) if no forces act upon a body then it will remain in the state it is; and (2) forces bring about the movement of a stationary body and change the movement of a body that is already moving.

One of the greatest scientists the world has known, Isaac Newton (see page 68), summed up this experiment in what is known as his first law of motion:

> *A body will remain at rest or continue in uniform motion in a straight line unless it is acted upon by a force.* ('Uniform' here means 'at the same speed'.)

The property of an object to remain in its present state, whether still or moving, is called

▼ **A test to see how a car and** its dummy passengers stand up to a crash. The car stops suddenly, but the passengers carry on moving forwards because of their inertia.

► **In snooker, players use**
Newton's laws of motion when
they change the direction and
speed of the balls. They apply an
external force with their cue, and
rely on inertia and momentum to
pot the balls in the pockets.

▼ **Firing muskets in a**
reenactment of a battle of the
English Civil War. The guns recoil,
or jolt backwards, when fired. This
follows Newton's third law of
motion. The force (action) of the
gases shooting forwards out of the
muzzle is accompanied by an
equal force (reaction) in the
opposite direction. This is the
recoil.

its inertia. It is a basic property of
matter. And the greater the mass
(see page 74) of an object, the
more inertia it has, and the more
difficult it is to get it moving or to
change its movement.

You can feel the effects of
inertia in a lift. When the lift
starts to go down, you feel light
on your feet and you feel your
stomach move upwards. You get
this feeling because your stomach tries to stay where it is as a
result of its inertia. Again, when

the lift stops on the lower floor,
you feel heavy on your feet, and
feel your stomach dropping. Its
inertia tries to keep it moving.

Momentum

Newton came up with two other
laws to describe the condition of
motion. His second law concerns
momentum, which is the mass
of an object multiplied by its
velocity, or speed, in a certain

direction. The law states that:

The rate of change of momentum of a body is equal to the applied force and takes place in the direction of that force.

In simple terms this law means that an object speeds up or slows down according to the size of the force acting upon it.

Action, Reaction

Newton's third law of motion is perhaps the best known of the three. It states that:

To every action, there is an equal and opposite reaction.

This means that whenever you apply a force (action) to an object, then that object applies a force back (reaction) in the opposite direction.

This law explains why you can easily fall into the water when you are stepping out of a small boat. The force you exert when you step forwards (action) out of the boat is accompanied by an equal force (reaction) in the opposite direction, which makes the boat move backwards. And if the boat isn't tied up, you will step out – into the water!

The force of reaction explains why a gun recoils, or moves backwards, when the bullet moves forwards. It also explains how rocket and jet engines work. They are called reaction engines.

Friction

In the everyday world, whenever something moves, forces are at work which try to stop it moving. The main force is friction, which means the rubbing of one thing against another.

If you try to roll a tennis ball over a thick carpet, it won't roll very far. The friction between the rough ball and the rough carpet is very high, and the force it sets up soon stops the ball.

But if you roll a ping-pong ball over a vinyl floor, it will travel much farther. Both the ball and the floor surface are smooth, so they do not rub together much. Friction is low.

Machines use up a lot of energy in overcoming the friction of their moving parts. To help reduce friction, the moving parts are oiled, or lubricated. The oil coats the parts and stops them rubbing.

Friction may be a nuisance, but it is also necessary. If there were no friction between your feet and the ground, you would slip whenever you tried to walk! This is what happens when you try to walk on ice, which has low friction.

The friction between things rubbing together gives out energy as heat. Rub your hands together and you will notice how hot they get. Friction with the air makes returning shuttles glow red-hot. They have a heat shield to protect the astronauts inside.

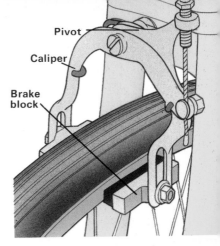

▲ **The caliper brakes on a bike** work by friction. When you apply the brakes, the scissor action of the calipers forces the brake blocks against the rim of the wheel. The friction produced makes the wheel slow down.

▼ **Toboggans slide easily on** snow partly because it has very low friction, but also for another reason. The children's weight exerts pressure on the snow and causes it to melt a little. The water has even lower friction.

Round and Round

▲ **On this corkscrew roller-**coaster, forces produced by the car's motion keep it on the track.

If you tie a ball on a piece of string and whirl it round your head, what happens? As long as you keep pulling on the string, the ball will travel round in a circle. It moves in this way because of the two forces that are acting on it.

The first force on the ball is produced by the movement of your arm. It acts in the forwards direction. The second force is caused by the tension in the string and your arm which prevents the ball flying off into the air. It acts at right-angles to

the first force, and is called a centripetal ('towards the centre') force.

The combination of these two forces produces a net, or resultant, force that leads to the ball moving in a circle. The same principle applies to any object following a circular path, such as the Moon around the Earth and the Earth around the Sun. A car going round a corner follows a curved path because of centripetal force from the friction of its tyres on the road. On a wet surface, the centripetal force is reduced and the car may skid.

People frequently talk about centrifugal ('away from the centre') force to explain the circular motion of the ball, but this is inaccurate. It is the inward, centripetal force produced by your pull which keeps the ball moving in a circle. Without it, the ball would fly off, not because it is being pulled away by a centrifugal force, but because it is propelled forwards by your arm.

In a spin

As the Earth travels through space around the Sun, it also spins on its axis. The direction in space in which the axis points never changes. This is due to its angular momentum – the momentum (see page 67) due to its rotation. The angular momentum of a spinning body remains constant if there are no other forces acting on it.

This principle is called the conservation of angular momentum. It can explain, for example, why ice skaters can make themselves spin faster when they draw in their arms. It also explains the unusual behaviour of the gyroscope. Once a gyroscope is set spinning, its axis always points in the same direction. This is why it can be used as a compass, in the gyrocompass. The gyrocompass is a more accurate direction-finding instrument than the ordinary magnetic compass as it is not affected by movement of its mounting or by the presence of any local magnetic fields.

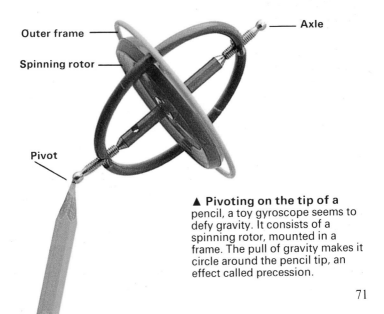

Outer frame

Spinning rotor

Axle

Pivot

▲ **Pivoting on the tip of a** pencil, a toy gyroscope seems to defy gravity. It consists of a spinning rotor, mounted in a frame. The pull of gravity makes it circle around the pencil tip, an effect called precession.

Gravity

If you drop something, it always falls. This happens because the Earth is pulling it. The Earth pulls everything on and around it with a force we call gravity. Gravity keeps our feet on the ground, the water in the oceans, the air in the sky, and the Moon in the heavens. Isaac Newton (see page 68) was the first person to understand the nature of gravity.

But gravity is not just a property of the Earth. Gravity is one of the great forces of the universe that literally holds all the bodies in the universe in place. Just as the Earth's gravity holds the Moon in place, so the Sun's gravity holds the Earth and all the other planets in the Solar System in their places. Gravity is a property all matter has to a lesser or greater degree. The more massive a body is, the greater is its gravitational pull.

The planet Jupiter, for example, is much more massive than the Earth, and has over $2\frac{1}{2}$ times the Earth's pull. On the other hand, the Moon has much less mass than the Earth, and has only one-sixth of the pull. If you can high-jump 1 metre on Earth, you could clear only about 40 cm on Jupiter. But on the Moon you could high-jump 6 metres!

▼ **Free-fall parachutists** linking in a spectacular formation. They accelerate to a certain speed until the air resistance on their bodies equals the pull of gravity.

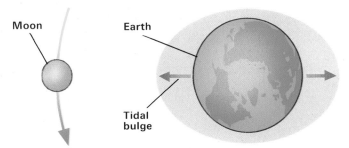

Even though the Moon's gravity is so low, it still reaches out to the Earth. We see its effect at the seaside with the rise and fall of the tides (above).

▲ **When the Moon is overhead**, its gravity pulls the water of the oceans towards it, and we have a high tide. When the Moon passes by, the bulge of water goes with it, and the sea level falls.

Falling bodies

If you dropped a ping-pong ball and a golf ball from the same height, which would hit the ground first? You might think the golf ball would, because it is heavier. But you would be wrong. They would both hit the ground together – try it and see.

Indeed any object will fall to the ground at the same rate, as long as we ignore air resistance. This follows from Newton's second law (see page 68), when it is applied to the gravitational force on a body. (In fact, air resistance can make a big difference, as can be seen by the way a parachute falls slowly rather than plummeting to the ground.)

GALILEO

An Italian scientist, Galileo (1564–1642), was one of the first to study the motion of bodies scientifically. He is said to have carried out experiments on falling bodies by dropping them from the Leaning Tower of Pisa. He is also well known for his study of the pendulum and, as an astronomer, for building the first practical refracting telescope in 1610. Using this, he saw mountains on the Moon, phases of Venus and four of Jupiter's moons.

Black holes

The gravitational pull of the Sun and other stars is colossal. The Sun's gravity, for example, can keep hold of the tiny planet Pluto over a distance of 6000 million kilometres.

But the heavenly body with the most powerful gravity of all is a black hole. This is what remains when a huge star dies. At the end of its life, it collapses with such force because of its high gravity that it eventually crushes itself out of existence. It just leaves a region of space with incredibly high gravity. This region will swallow anything nearby – even light. That is why we call it a black hole.

Weight

Every object on Earth is pulled to the ground by gravity. The more mass the object has, the greater is the pull, or force, it experiences. We call the force that an object experiences due to gravity its weight.

Note that mass and weight are quite different. Mass means the amount of matter in the body, which never changes. Weight is the force the body experiences due to gravity. It depends both on the mass and the force of gravity.

We measure the mass of an object in grams, kilograms (thousands of grams), and tonnes (thousands of kilograms).

Strictly speaking, we should measure weight, which is a force, in terms of units of force, called newtons (after Isaac Newton). But in everyday life we use grams, kilograms and tonnes as units of weight.

We can illustrate the difference between mass and weight in the following way. An astronaut with a mass of 84 kilograms weighs about 840 newtons on Earth. On the Moon, the astronaut's mass would still be 84 kilograms, but he or she would weigh just 140 newtons, because the Moon's gravity is only one-sixth that of Earth.

In balance

We measure the weight of things by means of scales. Simple scales were in use in ancient Egypt, and the same type is still in use in

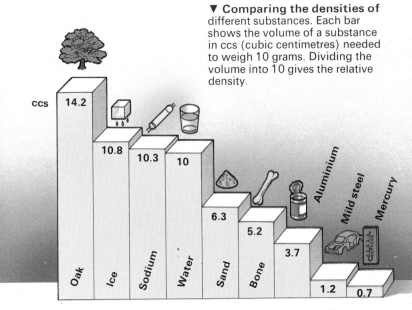

▼ Comparing the densities of different substances. Each bar shows the volume of a substance in ccs (cubic centimetres) needed to weigh 10 grams. Dividing the volume into 10 gives the relative density.

ccs

Oak	Ice	Sodium	Water	Sand	Bone	Aluminium	Mild steel	Mercury
14.2	10.8	10.3	10	6.3	5.2	3.7	1.2	0.7

many countries today. It depends on the principle of balance. It has two weighing pans, hanging one each side of an arm that is suspended in the middle. The material to be weighed is placed in one pan, then weights are placed in the other. When the arm is level, the weight of the material equals that of the weights.

The same principle is used in some of the accurate balances used for weighing in scientific laboratories. They can measure down to the nearest one ten-thousandth of a gram.

◀ **Simple scales are used in** street markets everywhere, as here in China. This one uses the stretching of a spring as a means of measurement.

Most household scales have a single pan. They work by means of a spring, which changes length according to the weight.

Density
We often talk about some things being heavier than others, such as lead being heavier than wood. But this is not an accurate statement – after all, 1 kilogram of lead weighs exactly the same as 1 kilogram of wood. So we should really say that lead is *denser* than wood. In other words, it has more mass for the same volume. Densities are often measured in units such as grams per cubic centimetre. But often they are expressed as relative densities – their density compared with the density of water.

Satellites

The Earth's gravity is very powerful. If you throw a ball up in the air, it will travel a little way and then come down. If you fire a bullet from a gun up in the air, it will travel faster and go farther before falling back to Earth.

In other words, the faster you propel an object, the further it will go before gravity pulls it back to Earth. You are starting to beat gravity by speed.

Imagine that you could propel something from the Earth parallel to the ground with a speed of 28,000 kilometres per hour. It would travel so fast that gravity would pull it back only very little. During the time it would take to fall 1 metre, say, it would have travelled so far that the surface of the Earth would have curved away 1 metre. In other words, it would stay the same height above the surface.

By using powerful rockets, we can propel objects at speeds of 28,000 kilometres per hour around the Earth so that they stay up. We must shoot them above the atmosphere into space, otherwise air resistance would

▶ **To travel in space a body** must overcome gravity. If it is launched at too low a speed, then it will soon return (top). If it is launched at 28,000 km/h, it will go into orbit (middle). If it is launched at 40,000 km/h, it will be able to escape from Earth completely (bottom).

Sub-orbital velocity

Orbital velocity

Escape velocity

slow them down. We call these objects satellites, and the paths they take in space their orbits. The Russian spacecraft *Sputnik 1* was the first satellite to be launched into orbit on 4 October 1957.

Most satellites circle in orbits of a few hundred kilometres high. But some circle in an orbit nearly 36,000 kilometres high. This is an interesting orbit in which a satellite takes exactly 24 hours to circle the Earth. In other words, it keeps pace with the Earth itself and appears to be fixed in the sky. This makes communicating with it from the ground very much simpler.

Weightlessness

We have noted that things in orbit are actually falling towards the Earth, but in practice not getting any nearer.

Astronauts in their spacecraft are in this state too, falling along with everything else. We call this state free fall. Popular names for it are weightlessness and zero-g (no gravity), because gravity seems absent and bodies appear to have no weight.

▼ **An American astronaut** flying the Manned Manoeuvring Unit. He appears to be weightless but gravity is still acting, stopping him flying off into deep space.

Energy

The world is full of energy in one form or another. The Sun beats down on the Earth, bringing us energy in the form of heat. A flick of the light switch allows electrical energy to flow into the light bulb, which gives out energy as light. When coal burns, it undergoes a chemical reaction that releases chemical energy in the form of heat and light. And in a battery, chemical energy is turned into electricity.

Kinetic and potential energy

One thing energy does is cause motion. When you ride a bicycle, you use energy to push the pedals. This makes the bicycle move. The energy released when a car engine burns petrol is used to make the car move. When an object is moving, we say it has kinetic energy.

A ball resting on the edge of a table would not appear to have

▼ **A hydroelectric power** scheme harnesses the potential energy in the water stored in the reservoir behind a dam. The potential energy transforms into kinetic energy when the water flows down into the turbines. The turbines acquire mechanical energy as they spin, which the generators transform into electrical energy.

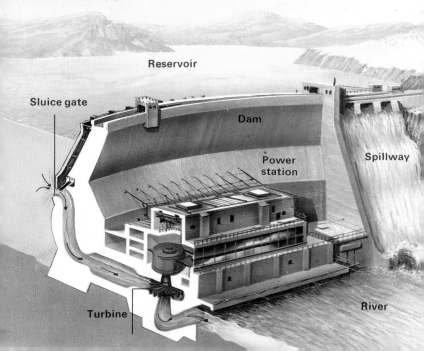

Reservoir

Sluice gate

Dam

Power station

Spillway

Turbine

River

▶ **An engineer uses** chemical energy to cut through a steel girder. He is cutting with an oxyacetylene torch, which burns a mixture of oxygen and acetylene (ethyne). The energy is released not only as heat but also as dazzling light.

any energy, but it has. It has the energy of position, called potential energy. If it is touched, it will fall from the table and start moving downwards. The potential energy has been changed, or converted into kinetic energy.

Conversion and conservation

The conversion of one form of energy into another happens all the time, as in the examples given above. Another good example of energy conversion is in a hydro-electric power plant.

The stored water behind the dam represents a store of potential energy. When the water flows down into the turbines, the potential changes to kinetic energy. The water spins the turbines, which now have mechanical energy. The turbines in turn spin the generators, which change the mechanical into electrical energy. Transmission lines then carry the electrical energy to homes and factories, where other conversions take place.

The electricity is turned into heat energy in electric heaters and light energy in electric lights. It is turned into mechanical energy by electric motors, to power all kinds of machines, from lawn mowers to workshop lathes and express trains.

If you could work out what energy was stored in the water to begin with, and what energy was produced at the end, you would find that they would be the same. This obeys the law of the conservation of energy, that energy can be neither created nor destroyed.

Work, Energy and Machines

Work and energy are both measured in the same units – joules. This unit is named after the British scientist James Prescott Joule, who first demonstrated in the 1800s that energy and work are really one and the same thing. One joule is the work done when a force (the use of energy) of 1 newton acts through a distance of 1 metre.

One definition of energy is 'the ability to perform work'. Energy only becomes useful to us when we can put it to practical use. We do this in machines. A machine can be defined as a device that performs work.

We use thousands of different machines in the modern world, from bicycles to combine harvesters. They are each made up of all kinds of parts, or components – wheels, gears, bearings, shafts, pistons, springs, rods – which work together in a complicated way to do a certain job. But the action of the separate parts is often very simple. They act as simple machines.

SIMPLE MACHINES

Some machines hardly seem like machines at all. They are the simple machines, such as the lever, inclined plane, wedge, screw and wheel and axle. They are not only able to apply forces, they can also magnify them. A small force (effort) applied to one part of the machine, can produce a larger force (load) at another. This is as a result of making the effort move farther than the load.

A lever is the simplest machine. An example is a seesaw, a plank that can turn about a pivot, or fulcrum, at its centre. A girl can balance a man twice her weight by sitting twice as far away from the pivot, proving the 'law of the lever' first worked out by Archimedes (see page 19). The law states that effort times effort arm (distance from pivot) equals load times load arm.

Other simple machines such as those shown here have similar useful effects. They are all good ways of magnifying force.

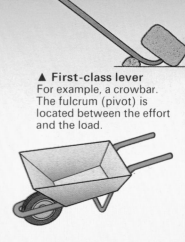

▲ **First-class lever**
For example, a crowbar. The fulcrum (pivot) is located between the effort and the load.

▲ **Second-class lever**
For example, a wheelbarrow. The fulcrum is at one end, and the load is between it and the effort.

▶ **Gears are a very useful** mechanical device for transmitting motion in machines. They can also modify that motion and bring about a change of speed. This is done by bringing together gear wheels with different numbers of teeth. In the example, the small gear wheel has 8 teeth, the large one 16. This means that the large gear will drive the small one round at twice its speed. On the other hand, the small one will drive the large one at only half its speed.

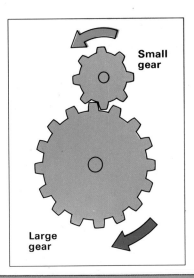

Small gear

Large gear

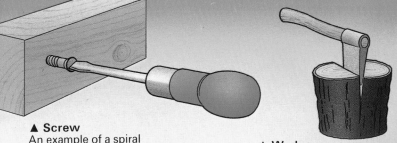

▲ **Screw**
An example of a spiral inclined plane. Effort applied to turn the head once advances the screw by the distance of one thread into the wood.

▲ **Wedge**
For example, an axe. Effort applied downwards exerts forces sideways to split the wood.

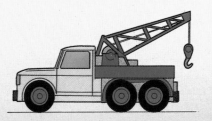

◀ **Wheel and axle**
For example, a winch. Turning the handle in a large circle makes the axle turn in a small one.

◀ **An oil production** platform in the North Sea, with gas flaring. The gas comes up from the undersea wells with the oil and is separated on the platform. Offshore oil fields are now being tapped in many parts of the world, requiring a range of new technologies for drilling and production in deep and stormy waters.

Sources of Energy

Our modern world uses up vast amounts of energy to light and heat our homes, to power the machines that produce our goods, to drive our cars, buses, trains and trucks, and for many more things besides.

Three main sources produce the energy we need: oil, coal and natural gas. Between them they provide over 85 per cent of the energy the world uses. The other energy mostly comes from hydroelectricity and · nuclear power stations. Hydroelectricity harnesses the energy in flowing water (see page 78). Nuclear power stations make use of nuclear energy, the energy given out when atoms split (see page 33). Supplies of the three main sources of energy will one day run out (see page 84).

Fossil Fuels

Oil, coal and natural gas are known as fossil fuels because they are the remains of plants and animals that once lived on Earth. Coal is the remains of plants – giant tree ferns and horsetails – that lived during the Carboniferous Period of Earth history, about 300 million years ago. Oil and gas are what is left of simple plant and animal life that lived in ancient seas.

From oil, or petroleum ('rock oil') as it is properly called, we obtain our most important fuels: petrol, kerosene and diesel (gas) oil. These are produced by processing petroleum at a refinery (see page 60). Both oil and natural gas are extracted from the ground by drilling.

Natural gas is a mixture of several gases, mainly methane. It also contains some propane and butane. By compressing these two gases they can easily be turned into liquid, which makes them a useful portable fuel supply, as bottled gas. Releasing the pressure turns them back into gas. The substances that make up oil and the gases in natural gas are mostly hydrocarbons (see page 55), compounds of hydrogen and oxygen only.

Coal contains mainly carbon. The best coal, called anthracite, is nearly all carbon. When coal is heated without contact with the air (a process called destructive distillation), it turns into another useful fuel: coke. This is widely used in blast furnaces (see page 108). The gas given off during coke-making is also a valuable industrial fuel. A similar process was once employed at city gasworks to produce coal gas, or town gas. This was the heating and lighting gas used in homes before natural gas.

THE GREENHOUSE EFFECT

The Earth receives most of its energy from the Sun. Solar energy passes through the atmosphere and warms the Earth. The Earth radiates a lot of this energy back into space. But the presence of heavy carbon dioxide in the atmosphere prevents some of this heat escaping, just as the glass stops heat escaping from a greenhouse. This atmospheric 'greenhouse effect' will accelerate if we continue to pump in more carbon dioxide. World temperatures will rise markedly. Climates will change and the polar ice caps will melt, causing widespread flooding in coastal regions.

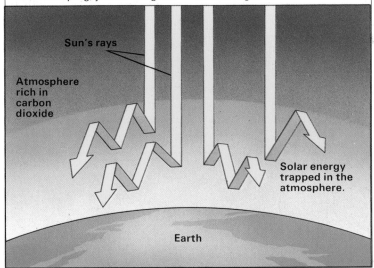

Sun's rays

Atmosphere rich in carbon dioxide

Solar energy trapped in the atmosphere.

Earth

Alternative energy

To obtain the energy it needs, the world has to burn colossal amounts of oil, coal and natural gas. Up to 6000 million litres of oil alone are used up every day. At this rate the world will run out of oil in less than 50 years. The position with natural gas is about the same. Only coal is plentiful enough to last for a long time – perhaps 200 years.

Scientists and engineers are making great efforts to develop alternative sources of energy so that we can make the fossil fuels last as long as possible. Some think that we should use more nuclear power (see pages 30 to 35). But this is not good for the environment because of the deadly radioactive wastes nuclear plants produce. The most sensible way is to use the abundant energy provided by nature, such as the wind, flowing water and sunlight, which are free and everlasting.

We are already tapping the energy of flowing water in hydroelectric schemes (see page 78). We are also starting to tap the ebb and flow of the tides. In places the difference between high and low tides is as much as 12 metres, and this is enough to

▼ **Harnessing wind energy on** a large scale on a 'wind farm' at Altamont Pass, California. It consists of some 300 propeller-type turbines, which feed electric power into the local grid system.

drive water turbines. A tidal power plant at the mouth of the River Rance in the northwest of France has been operating since 1966.

Solar energy is also now being widely used for heating water in the home. Glass-covered solar panels, rather like miniature greenhouses, are used to trap the heat. Large-scale solar energy schemes have been set up in sunny climates, such as California, USA, to produce electric power. They consist of a huge array of mirrors that reflect the Sun's heat on to a tower. There it boils water into steam to drive turbines that produce power.

The Sun's energy is also harnessed indirectly in wind-power

▲ **This futuristic car is** powered by solar energy. The energy from the Suns' rays is collected by the panels at the back.

schemes. These use propeller-type turbines, which are the modern equivalent of the old windmills.

Several countries are also tapping the Earth's heat, or geothermal energy. This is held deep down in underground rocks, especially in volcanic regions. In such regions, water often makes its way down into the rocks and is heated. It may bubble its way back up to the surface, forming hot springs, or shoot out of the ground as a geyser, or fountain of steam and water.

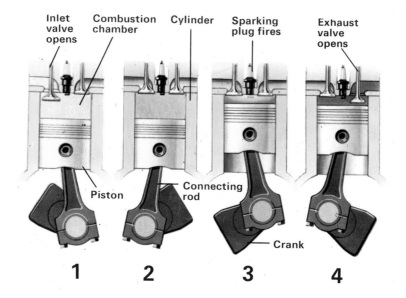

Inlet valve opens   Combustion chamber   Cylinder   Sparking plug fires   Exhaust valve opens

Piston   Connecting rod   Crank

1   **2**   **3**   **4**

Engines

Engines are machines that change energy into mechanical motion. A car engine, for example, changes the energy in petrol into mechanical motion to turn the wheels of the car. A lorry engine changes the energy in diesel oil into mechanical motion to turn the lorry's wheels.

In petrol and diesel engines, energy is released by burning the fuel inside a confined space. They are called internal combustion engines. Combustion (burning) takes place inside a number of cylinders containing pistons. The hot gases given off force the pistons down the cylinders to produce power. Rods transfer the motion of the pistons to a shaft (the crankshaft) and spin it

▲ The four-stroke cycle of a petrol engine, showing the actions of the pistons and the valves.

round. In a vehicle, the transmission system then transmits, or carries, the rotary motion of the crankshaft to the driving wheels.

Four-stroke engines

In a petrol engine, certain actions take place in each cylinder according to a regular cycle. This ensures that the engine keeps running. The cycle is repeated every four strokes (up or down movements) of the piston.

The stages of the four-stroke cycle are illustrated at the top of the page. The cycle begins when a valve opens at the top of the cylinder. A mixture of petrol and air is drawn in as the piston moves down on its first stroke (induction, or intake). The valve shuts as the piston moves up on its second stroke (compression), compressing the fuel mixture. At the top of this stroke, the fuel is ignited by an electric spark. The hot gases produced expand and force the piston down on its third stroke (power). When the piston goes back up again on its fourth and final stroke in the cycle (exhaust), it pushes the burnt gases out of the cylinder through another valve. Then this valve closes and the cycle is repeated.

The diesel engine works on a slightly different four-stroke cycle, in which the fuel ignites when it is injected into hot, highly compressed air. That is why the diesel is called a compression-ignition engine.

Rotary motion
Engines that use pistons moving up and down in cylinders are called reciprocating engines. They must have a crankshaft (a specially shaped metal shaft) to produce useful rotary (turning) motion. Other engines produce rotary motion directly. They are called turbines and work rather like a windmill and spin when a fluid flows through them.

The most powerful land-based turbines use steam to spin the turbine wheels. Water turbines are used in hydroelectric power plants. Jet planes use a form of gas turbine in their engines. The turbines spin the compressors that force air into the engines.

▼ A cutaway picture of a simple jet engine. Air enters the engine at the air intake (1). It is compressed in two stages by the compressors (2, 3) and enters the combustion chamber (4), where fuel is sprayed in and ignited. The gases produced shoot out of the rear as a propulsive jet, spinning the turbine (5) as they do so. This drives the compressors.

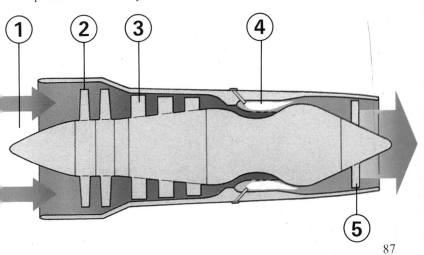

Light and Radiation

Sight is perhaps the most important of our five senses. We see because our eyes are sensitive to light. During the day that light comes from the Sun. Light rays are one way in which the Sun gives out its energy. But it gives off energy in radiation (rays) of many other kinds as well – gamma rays, X-rays, ultraviolet rays, infrared rays, microwaves and radio waves. Light and all these other rays are basically very similar. They belong to a family of rays called electromagnetic radiation. This radiation is so called because it consists of vibrating electric and magnetic waves. Unlike sound waves and water waves, electromagnetic waves need no medium (a solid, liquid or gas) to travel in. That is why they can reach the Earth across space.

The various kinds of rays differ from one another in their wavelength, the distance between the crest of one wave and the next. Light itself is not a single wavelength, but a mixture of wavelengths. When we split it up into its separate wavelengths, we find that each is a separate colour – one of the colours of the rainbow.

Light travels through some substances, such as glass and water, and is reflected by others, such as shiny metal. When it passes into glass or water, it is bent, or refracted. We make many useful instruments by using the properties of reflection and refraction. They include telescopes, microscopes and cameras. Telescopes enable us to see objects in the heavens whose light has taken billions of years to reach us. Microscopes enable us to see objects thousands of times smaller than a pinhead. Cameras can capture images forever, and even 'stop' the motion of a bullet in mid-air.

▶ **A rainbow appears in a** stormy sky in New Mexico, USA, the colours caused by raindrops splitting up sunlight into different wavelengths. In the foreground are five of the 27 aerials of the Very Large Array radio telescope, which tunes in to radio waves from outer space.

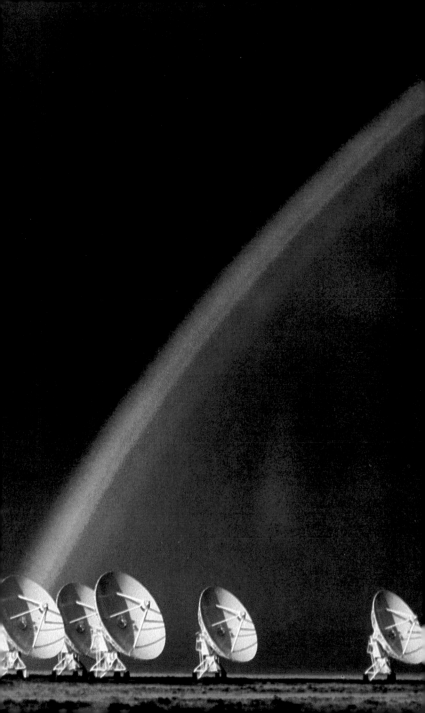

The Electromagnetic Spectrum

▼ **The diagram shows the whole range of electromagnetic** wavelengths, spread out in a band, or spectrum, going left to right from the short-wavelength gamma rays to the long-wavelength radio waves. Note how we have been able to utilize most of the spectrum in one way or another.

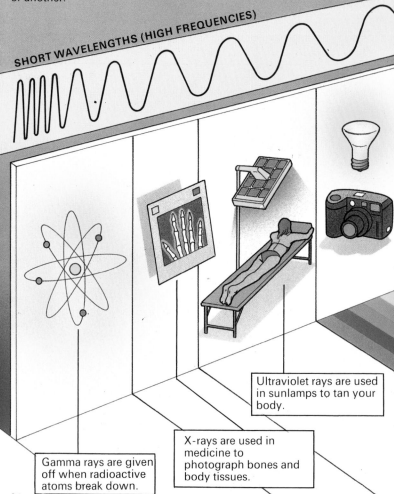

SHORT WAVELENGTHS (HIGH FREQUENCIES)

Ultraviolet rays are used in sunlamps to tan your body.

X-rays are used in medicine to photograph bones and body tissues.

Gamma rays are given off when radioactive atoms break down.

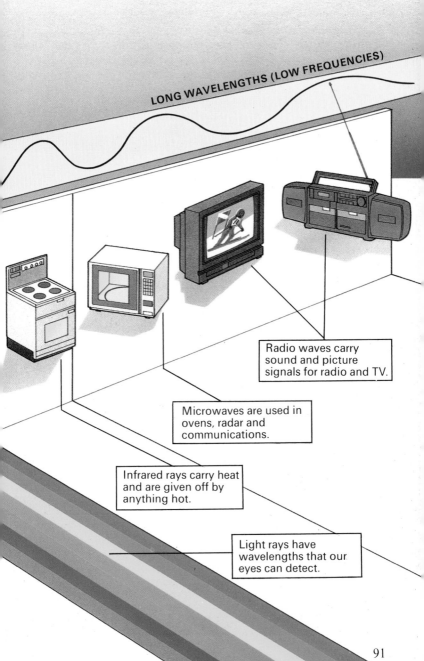

LONG WAVELENGTHS (LOW FREQUENCIES)

Radio waves carry sound and picture signals for radio and TV.

Microwaves are used in ovens, radar and communications.

Infrared rays carry heat and are given off by anything hot.

Light rays have wavelengths that our eyes can detect.

The short and the long

Electromagnetic waves are electric and magnetic disturbances, or ripples in space. If you could see them, they would be much like the waves that form on water when you throw a stone into a pond. These are called transverse waves.

The wavelengths of electromagnetic waves vary enormously. The shortest gamma rays have a wavelength of less than one million-millionth of a metre. Scientists call this measurement 1 picometre.

The wavelengths get longer through X-rays, ultraviolet, visible, infrared, microwaves and radio. Radio waves have the longest wavelengths – up to thousands of kilometres.

But whatever their wavelength, all electromagnetic waves travel at the same speed – the speed of light. We don't normally think of light having a speed, but as something that travels instantly. In our everyday world it more or less does, for its speed is very nearly 300,000 kilometres per second. This means it could travel nearly eight times around the world in one second! Nothing can travel faster than light. It is the fastest thing in the universe.

▼ **A patient undergoing** investigation on an X-ray scanner. This machine probes internal organs by a technique called tomography. It produces an X-ray image of a thin 'slice' of the body.

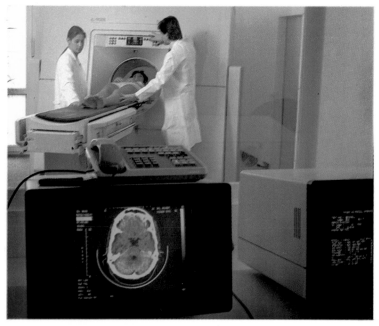

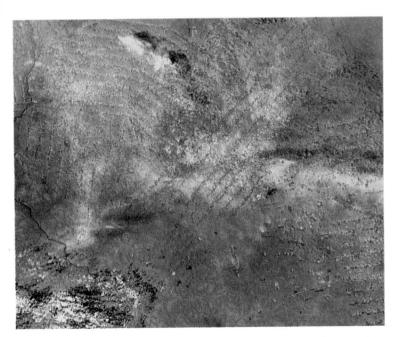

Frequency

We often describe radio waves, not by their wavelength, but by their frequency. The frequency is the number of waves that pass a certain point in a second. It is inversely proportional to the wavelength. This means that when the waves are long, the frequency is low (few waves go past); and when the waves are short, the frequency is high (more waves go past).

The frequency of a wave is measured in units called hertz: 1 hertz is the frequency when one complete wave passes a certain point in one second. (It used to be called 1 cycle per second.) The unit is named after a German physicist, Heinrich Hertz, who in

▲ **The satellite Landsat took** this picture of the Amazon rainforest at invisible wavelengths to highlight forest destruction.

1889 was the first to show the existence of electromagnetic waves.

Because long radio waves have the longest wavelengths, they have the lowest frequency, of about 1 kilohertz (1 KHz), or 1000 hertz. VHF (Very High Frequency) radio waves have a frequency of the order of 100 megahertz (1 MHz), or millions of hertz. The microwaves used in microwave ovens and in radar have frequencies of several gigahertz (GHz), or thousands of millions of hertz.

Looking at Light

One basic property of light we can all observe is that it travels in straight lines. This explains why we can't see round corners and why objects cast sharp shadows in strong sunlight. When light falls on a surface, a number of things can happen. It may bounce back, or be reflected – this is what happens with mirrors. Or it may pass through. We say that substances which let most of the light pass through are transparent. And those that let no light through are opaque. When light passes into a transparent substance, it bends, or is refracted (see page 98).

Light may also be absorbed. The absorption of different wavelengths by an object determines the colour we see (see page 104).

Isaac Newton (see page 68) was one of the first people to study light scientifically. He believed that it was made up of a stream of particles, which he called corpuscles. This can explain why light travels straight and how it bounces off mirrors. But it cannot explain refraction or two other phenomena – diffraction and interference.

Diffraction is the spreading out of light slightly after it passes

◄ **Shafts of light** penetrate the gloom of a dense forest. Notice how straight they are, demonstrating that light rays travel in straight lines.

through a narrow slit. Interference sometimes occurs when two beams of light meet. It produces a series of light and dark bands.

Interference is an effect which can easily be explained by wave motion. We can observe it between two ripples on a pond. In places where the crests (high points) of the two waves meet, they combine to make a higher one. In places where a crest and a trough (low point) meet, they cancel out.

And this is what happens in light. When the waves of two beams are in step (their crests coincide), they produce brightness. When the waves are out of

▲ **Blowing bubbles and rainbows.** Little rainbows appear in bubbles because of interference between light reflected from the top and bottom surfaces of the soap film. This splits light into a spectrum.

step (crest meets trough), they cancel out, producing darkness.

Waves *and* particles
The Scottish physicist James Clerk Maxwell first showed that light is indeed a wave – an electromagnetic wave – in the 1860s. Newton's idea of light particles seemed dead. But early this century it was revived in order to explain how light is emitted and absorbed in atoms.

The German physicist Max Planck showed in his quantum theory that light (and other radiation) is emitted and absorbed in little 'packets', or quanta. Each quantum behaves like a tiny particle, which we call a photon. So it appears that in different circumstances, light can behave as a wave or as a particle.

Reflection

We see things because they reflect the light that falls on them back into our eyes. If they didn't reflect light, they would be invisible. When a surface is very smooth and shiny, it reflects most of the light that falls on it. It becomes a mirror. Most mirrors are sheets of glass with a silvery coating on the back.

The simplest mirror is the plane (flat) mirror, the ordinary kind we have at home. When you look in a plane mirror, you see a picture, or image, of yourself. But it is not a true picture. If you raise your right hand and wink your left eye, your image will raise its left hand and wink its right eye. In a mirror, left and right are reversed.

Your mirror image appears to stand as far back behind the mirror as you stand in front. But a mirror image does not really exist. We call it a virtual image.

Useful curves
Mirrors are also useful when they are curved. A shaving or make-up mirror is an example. When you look into the mirror, you notice that it magnifies your face, making shaving or making up easier. It is an example of a mirror that curves inwards, or a concave mirror. The image is again a virtual one.

If you step back a little way, you notice that the image you see

◄ **This trick fairground mirror** is made with a surface that curves in different directions. It produces a distorted image, often with amusing effects.

in the mirror turns upside-down. This image does exist, and will appear on a card screen held in the right place. It is a real image. Astronomical telescopes use concave mirrors to gather light and form real images of the heavens. It does not matter that they are upside-down. Isaac Newton built the first such reflecting telescope in 1668.

Mirrors that curve outwards towards you are also useful. They are called convex mirrors. In a convex mirror the image of an object is always virtual, smaller, and the right-way up. Convex mirrors have a wider field of view than an ordinary plane mirror. This means that they show more of the scene in front of them. They are often used for the rear-view mirrors of cars for this reason.

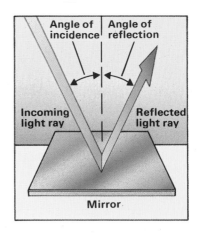

▲ **When light falls at an angle** on a plane (flat) mirror, it leaves at the same angle.

▼ **A 'spray' of optic fibres.** These are able to transmit light over long distances by reflection inside the fibres. They can be used to carry television pictures and telephone conversations.

Bending Light

When light is travelling through a transparent substance, such as air or water, it travels in a straight line. But when it travels from one transparent substance into another, it bends. We call the bending of light refraction. The reason light bends at the boundary of two substances is that it has a slightly different speed in each one.

Refraction explains why a stick that is half-in, half-out of the water looks broken. It explains why water in a bucket looks shallower than it is. It also accounts for mirages. You can see a mirage on the road on hot summer days. It looks as if there is a puddle of water on the road in the distance. Thirsty travellers in the desert often see pools of water, which also usually turn out to be mirages!

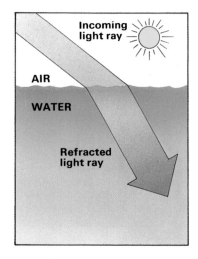

What they see is an image of the sky. Light coming from the sky is refracted by layers of air of different temperature and density. It curves into the travellers' eyes and appears to come from the ground ahead.

Lenses

We make use of the way light bends in many optical instruments, such as microscopes and cameras (see page 100). They are made using lenses, pieces of glass that have curved surfaces. Those that curve outwards (bulge in the middle) are called convex lenses. Those that curve inwards (go in at the middle) are called concave lenses.

◀ A 'broken' thermometer in a glass of water. The thermometer isn't broken of course; the effect is caused by refraction, the bending of light rays as they enter and leave the water.

◄ **Light travels faster in air** than it does in water. When it enters water, it slows down very slightly. This makes it change direction, causing the effect we call refraction. A similar thing happens when light enters glass, and this is how lenses work.

▼ **Lenses work by using** refraction at glass surfaces. By curving the surfaces in different ways, lenses can be made to converge light rays – bring them together (top); or diverge light rays – make them travel farther apart (bottom).

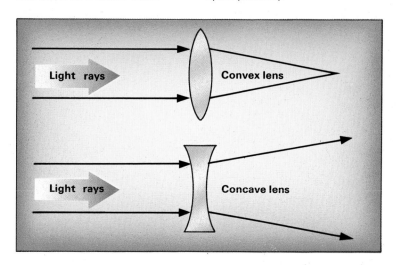

Convex lenses are the more useful of the two types. A simple magnifying glass is a convex lens. It produces a magnified, upright and virtual image when held close to the object being viewed. But the same lens will also produce a real image of an object some distance away.

One in the eye

Our eyes use a convex lens to produce a real image on the retina, a 'screen' at the rear of the eyeball. To focus on objects at different distances, it changes its curvature – the amount its surface curves. In some people, the lenses do not have the right curvature, and produce blurred images. But these people can correct their vision with external lenses. These may be spectacles or contact lenses.

The eyes of short-sighted people bring images to a focus in front of the retina. They need extra concave lenses to widen out, or diverge, the light rays entering their eyes so they focus properly on the retina. Long-sighted people, however, need convex lenses to bend in, or converge, the rays entering their eyes so that they form an image further forward, on the retina.

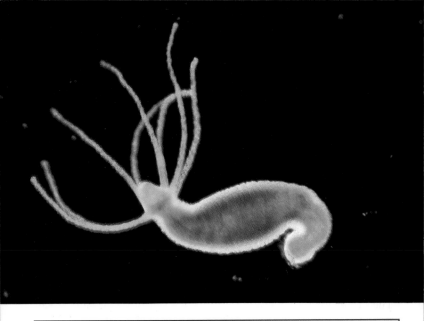

Optical Instruments

Optical instruments are those that use mirrors or lenses to bend light rays so as to produce images.

One simple instrument that uses mirrors is a submarine's periscope. This enables sailors under the water to see what is happening on the surface. It is essentially a tube containing a mirror at either end. They are fixed one above the other at an angle of 45°. Light rays enter the top mirror and are reflected through a right-angle down the tube to the bottom mirror. In between, they pass through lenses which magnify and focus the image. From the bottom mirror,

they are again reflected through a right-angle into the eyes of the viewer.

Camera and film
Photography – 'drawing with light' – is the world's favourite hobby. Photographs are 'drawn' using a camera, an optical instrument that imitates the eye. It uses a lens (actually a group of lenses) to form an image on a piece of film. This is coated with a light-sensitive layer, or emulsion, which contains silver salts. These change chemically when they are exposed to light. After exposure, they hold an invisible image of the light pattern that

◀ **Two optical instruments – a** microscope and a camera – were needed to record this image of a common resident in freshwater ponds, the hydra.

entered the lens. Developing, or chemical processing, turns the invisible image into a visible photograph.

A reflecting telescope uses mirrors rather than lenses to produce an image. A curved primary mirror gathers the light and reflects it on to another mirror, which reflects it into a viewing eyepiece.

Seeing near and far

The most useful optical instrument scientists use is the light microscope. This uses two sets of lenses to form a greatly magnified image of an object. The most powerful instruments can magnify up to about 2500 times.

(Electron microscopes, which use beams of electrons instead of light rays, can magnify even more – millions of times.)

A microscope has two sets of lenses – an objective lens (close to the object) and an eyepiece, or viewing lens. The objective forms a magnified image of the object, then the eyepiece magnifies that image further.

A refracting telescope works in much the same way, only its lenses are arranged so that they produce a magnified image of a very distant object rather than a very close one.

▼ **The light path through a** single-lens reflex camera. It uses multiple reflections by a mirror and a prism to produce an image in the viewfinder. In the viewing mode shown here, the mirror intercepts the light rays. It flips up when the shutter is released and allows light to fall on the film.

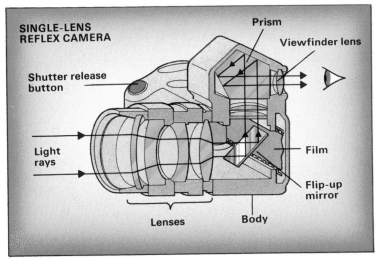

SINGLE-LENS REFLEX CAMERA

Prism

Viewfinder lens

Shutter release button

Light rays

Film

Flip-up mirror

Lenses

Body

Rainbow Colours

One of the most beautiful sights nature has to offer is the rainbow. This colourful arc appears in the sky when the Sun is behind you, and it is raining in front of you. Rainbows can also be seen near waterfalls. The major colours in the rainbow are violet, indigo, blue, green, yellow, orange and red.

We can produce a rainbow of colour artificially by passing light through a wedge-shaped chunk of glass called a prism (see below). The rainbow band we get is called a spectrum. The colours represent the various wavelengths that make up light (see pages 90–93). The violet rays have the shortest wavelength, while the red have the longest.

The spectrum comes about because each of the wavelengths is refracted to a different extent. They therefore separate out as they pass through the prism. The violet rays are refracted the most; the red rays the least.

The natural rainbow comes about because of refraction in the raindrops. Light is split up when it enters the drop, and is then reflected inside the raindrop, before being refracted even more as it comes out. Sometimes you can see a second rainbow

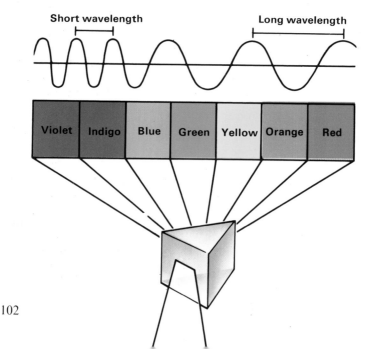

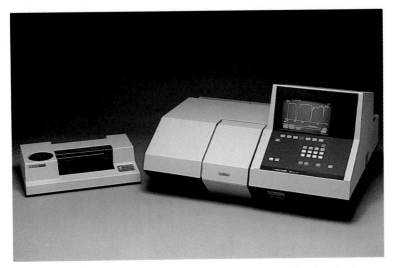

outside the first, with the colours reversed. This happens when the rays are reflected twice inside the raindrop.

The spectrometer

Scientists use an instrument called a spectrometer to study the light given off by glowing bodies. The instrument uses either a prism or a diffraction grating – a plate with many fine lines drawn on it – to produce a spectrum, and a telescope to examine it.

Astronomers find the spectrometer invaluable. By studying the spectrum of a star, they can find out all manner of things about it, such as what chemical

◄ When white light is passed through a glass prism, it does not emerge as white light but as a spectrum, or spread of different colours. These colours correspond to different wavelengths.

▲ The electronic spectro- meter plots spectra on a visual display. It is an invaluable tool of modern science.

elements it contains, whether it is moving towards or away from us, and how hot it is.

The most interesting thing about the spectrum of a star is that it has dark lines running across it. These dark lines come about in the following way. The star gives off light at all wavelengths, producing a spectrum of continuous colour. But when this light passes through the star's cooler outer atmosphere, the chemical elements present absorb certain wavelengths. This results in the dark lines appearing. From the positions of the lines astronomers can tell which chemical elements were responsible. This is how they know what stars are made of.

Mixing colours

We see colour because our eyes have special cells, called cones, which respond to different wavelengths of light. The cones are mainly sensitive to the wavelengths of blue, green and red. We call these colours the three primary colours of light.

Note that they cover the short, middle and long wavelength regions of the visible spectrum. When the eye receives equal amounts of these three colours, it thus receives all the wavelengths in white light and therefore sees white. In other words, when blue, green and red light are added together in equal proportions, the result is white.

All the other colours can be created by adding together two or three of the primary colours in different proportions. Adding red and green, for example, makes yellow; adding red and blue makes magenta; and adding green and blue makes cyan. Yellow, magenta and cyan are called the complementary, or

▲ **The primary colours of** light are red, blue and green. All three colours mixed together in equal proportions make white.

▲ **The primary colours in** printing inks are magenta, cyan and yellow. All three mixed in equal proportions make black.

◄ **When light falls** on an object, part is absorbed and part is reflected. When light falls on the leaves of a plant, only green light is reflected, and we see them as green. Other colours are absorbed.

secondary colours of light.

Colour television works on the principle of adding different proportions of the primary colours to create the whole colour range. If you look closely at a TV screen, you will see separate dots or stripes, coloured blue, green and red. They are materials called phosphors which give out coloured light when the electron beam strikes them (see page 156). From a distance the colours merge to create a properly coloured and meaningful image.

Subtracting colour

We see objects because they reflect light. Some objects reflect almost all the white light that falls on them, so we see them as white. But most objects absorb light as well. More specifically, they absorb different wavelengths of light. The wavelength they reflect gives them the colour we see. In other words, an object coloured red will reflect red but absorb all other colours.

Colour mixing in paint works on this principle of colour absorption, or subtraction. All the different colours can be obtained, as in light, by mixing three primary colours together. But the primary colours of paint are different – they are cyan, magenta and yellow. These three mixed together can produce any colour. If they are mixed together in equal proportions, they produce black.

▲ **All the colours in a** printed picture are produced by mixing just three colours – yellow, magenta and cyan – plus black. Four printing plates are used. The colours are printed one after the other, followed by black. The result is a true-to-life, full-colour image (see page 95).

Laser Light

The laser is one of the most exciting inventions of the century. Theodore Maiman built the first one in the US in 1960. The laser produces a very intense beam of light that can carry out eye operations, cut metal, play compact discs and transmit telephone calls. Obviously, it must be a very special kind of light.

A beam of ordinary white light is made up of many waves of different wavelengths vibrating in all directions. The waves, as it were, jostle one another and lose energy as a result.

Laser light is different. It is made up of light waves of exactly the same wavelength. These are exactly in step and travelling in exactly the same direction. This means that their crests and troughs occur at the same points. Because of this, the energy of each wave adds with the energy of all the other waves to produce an intense parallel beam of light, called coherent light.

▶ **A beam of ordinary light is** made up of many different wavelengths, or colours (left). Laser light is all one wavelength, and its waves are all in step (right).

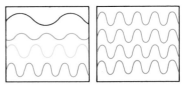

▼ **A ruby laser. The flashlamp** puts in energy, giving atoms in the crystal extra energy. They release this energy as radiation, which builds up as it is reflected between the mirrored ends. It emerges from one end as a very intense and parallel laser beam.

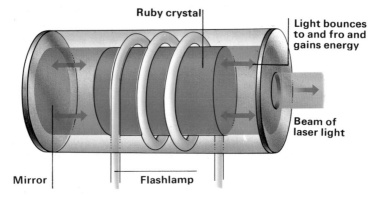

Ruby crystal

Light bounces to and fro and gains energy

Beam of laser light

Mirror

Flashlamp

LASER SURGERY

◀ **Lasers are invaluable in** eye surgery. They can be used to weld detached retinas back in place. (The retina is the eye's 'screen'.) They can also be used to puncture the eyeball to relieve pressure, which causes a condition called glaucoma.

▶ **Surgeons also** use lasers instead of scalpels to make incisions and to cut internal tissues. The great advantage of laser scalpels is that the beam heat-seals cut blood vessels, and helps reduce bleeding.

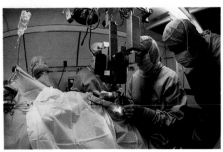

Stimulation

The word laser is coined from the name of the process by which laser light is produced. This is called 'light amplification by the stimulated emission of radiation'. It can take place in a crystal (see page 15) such as ruby; a gas, such as carbon dioxide; and a semiconductor (see page 146), such as gallium arsenide.

In a ruby laser (left), laser action is brought about by the atoms of the metallic element chromium that are present. When energy is injected into the ruby by a powerful flash lamp, the chromium atoms become 'excited', or gain energy. Some almost immediately emit this energy as a little flash of radiation of a specific wavelength.

When this radiation strikes other excited atoms, it stimulates them to emit radiation of the same wavelength and in step with it. The radiation builds up energy as it is reflected back and forth between the parallel, reflective ends of the laser. It emerges from the laser as a powerful parallel beam of pure light.

The ruby laser produces pulses of laser light. Most gas and semiconductor lasers are designed to work continuously, as they need to in the compact disc player, for example (see page 127).

Heat

Heat is a form of energy which we can easily sense. We feel the heat of hot water in a bath, the heat above a radiator and the heat from the Sun on a summer's day.

But what exactly is heat? It is the energy possessed by the particles of matter as they move. In other words it is a form of kinetic energy. The more energy that goes into a body, the faster its particles will move, and the hotter it will become. We measure the hotness of a body by its temperature, using a thermometer ('heat-measurer').

Heat travels. The heat from the hot water in the bath warms our bodies by conduction, or contact. The heat above a radiator reaches us through the air, by convection. The heat from the Sun reaches us by radiation: waves of energy travelling through space.

Heat, or the lack of it, plays an important part in our lives. We, and all warm-blooded creatures, need to keep our bodies at just the right temperature, otherwise we will die. The world as a whole relies on the heat from the Sun to stay alive. Without it, our world would be cold and lifeless.

We put heat to work in heat engines, such as the petrol and diesel engines that power our vehicles. They harness the heat energy given out by burning fuels and convert it into the energy of motion, to turn the driving wheels. We use other heat 'engines' to create cold – to extract heat from substances. The domestic refrigerator is the most familiar example.

▶ **Molten iron cascades from** a blast furnace. Heat is produced in the furnace by burning coke in a hot air blast. Temperatures of 1500°C are reached, more than enough to melt the iron.

Expansion and Contraction

If you heat up a cupful of milk for a drink, and then pour it back into the cup, you will find that you have some left over. This is because the milk expanded as it got hotter. If you blow up a balloon, and then put it in the refrigerator, you will find after a while that the balloon has got smaller. This is because the air inside contracted (took up less space) as it got cooler. And in general, things expand when they are heated and contract when they are cooled.

We can see why this happens when we look at matter on the atomic scale. When a substance is heated, its particles gain energy and vibrate more vigorously. They therefore take up more space. On the other hand, when a substance is cooled, its particles lose energy and vibrate less. They therefore take up less space, because they can pack together more closely.

All three forms of matter – solid, liquid and gas – tend to expand when they are heated and

▼ **A clinical thermometer**. It differs from an ordinary mercury thermometer by having a narrow constriction above the bulb. The mercury expands through this when the temperature is taken, but cannot flow back again until the thermometer is shaken.

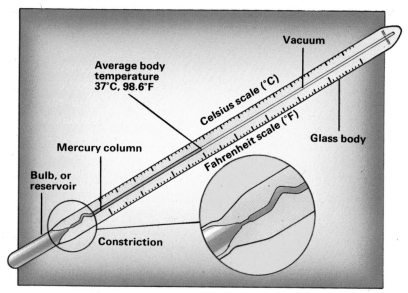

Average body temperature 37°C, 98.6°F

Celsius scale (°C)

Fahrenheit scale (°F)

Vacuum

Glass body

Mercury column

Bulb, or reservoir

Constriction

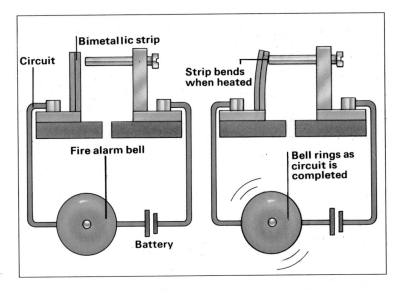

contract when cooled. Solids do not expand very much because their particles are bound rigidly together. Liquids expand more because their molecules can move more freely. Gases expand the most because their particles have little effect on one another.

Although solids expand and contract only a little, it is enough to cause problems for engineers building bridges, roads and railways. In building concrete roads, for example, gaps must be left at intervals to allow the concrete to expand, otherwise it would buckle in hot weather.

Different materials expand and contract by different amounts when heated and cooled. This property is put to good use in the thermostat, which can be used to regulate temperatures, or to create a fire

▲ **The two metals in a** bimetallic strip will expand by different amounts when they are heated, which makes it bend. This can be used to work a fire alarm.

alarm (see above).

The expansion of liquids is put to good use in making everyday thermometers to measure temperatures. They consist of a narrow glass tube containing a column of mercury or coloured alcohol. The position of the top of the mercury column on the thermometer scale indicates the temperature.

Most thermometers use a scale devised by a Swedish scientist, Anders Celsius. On this scale, water freezes at 0°C (Celsius) and boils at 100°C. Degrees Celsius are often referred to as degrees Centigrade instead.

111

Heat on the Move

When you make a cup of coffee with hot water and leave it, what happens? It gradually gets colder. The heat energy in the cup leaks away. It travels from the hot coffee to the cooler surroundings – the air and the table the cup rests on.

Heat always flows from a hot to a cold body; never the other way round. This is a simple statement of a basic law of thermodynamics, the science of 'moving heat'. Heat can flow from a hot to a cold body in one of three different ways – conduction, convection and radiation.

When you take a hot bath, your body warms up by conduction. The water passes on its heat to your body by contact. Its

▲ A hang glider soaring on a thermal, a rising current of warm air. This is a convection current in the atmosphere. The pilot will have to find several thermals if he is to stay aloft for any length of time.

particles are vibrating faster because it is hotter, and its particles make the particles of your skin vibrate faster too. So your body heats up. Metals are very good conductors of heat. But other substances, such as cork and glass wool, conduct heat poorly. We call them insulators.

A room with a hot radiator in it heats up by convection. The air by the radiator gets hotter and expands. As it expands, it becomes lighter, or less dense, than

▲ **During the day land heats** up more than water. At the seaside, this causes convection currents we call sea breezes.

▲ **During the night land loses** its heat more than water. At the seaside this causes convection currents we call land breezes.

the surrounding cold air. And so it rises. Cold air moves in to take its place, and that too is heated and rises in turn. In this way a current of warm air – a convection current – begins circulating around the room. On a broader scale, convection currents bring about the movements of air in the atmosphere which we call the winds.

The Sun's heat reaches us by radiation. It cannot do so by conduction or convection because there is nothing in space that can physically carry the heat. The radiation from the Sun is an electromagnetic radiation similar to light but with different wavelengths (see page 90). Its wavelengths are just longer than that of red light, and so it is called infrared radiation (infra means 'below'). Any hot body gives out infrared rays to a certain extent – even human bodies.

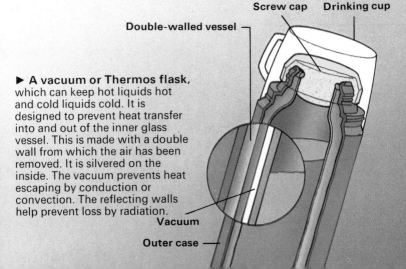

Screw cap   Drinking cup

Double-walled vessel

▶ **A vacuum or Thermos flask,** which can keep hot liquids hot and cold liquids cold. It is designed to prevent heat transfer into and out of the inner glass vessel. This is made with a double wall from which the air has been removed. It is silvered on the inside. The vacuum prevents heat escaping by conduction or convection. The reflecting walls help prevent loss by radiation.

Vacuum

Outer case

Keeping Cool

On a very hot day our bodies perspire to help keep us cool. The pores of the skin give off water. The water then evaporates, or turns into vapour, in the air. To do so, it must take in heat (known as latent, or 'hidden', heat). It takes this heat from the body, which is thereby cooled. The same principle is used to bring about cooling inside a refrigerator.

The refrigerator

In a refrigerator, a substance called a refrigerant is used which can readily change its state – from liquid to vapour and back again. Whenever a substance does this, it either absorbs or gives out latent heat.

The refrigerant circulates as a liquid up to a valve in the evaporator, around the freezing compartment of the refrigerator. As the liquid passes through the valve, it suddenly expands, because the pressure is much lower, and turns to vapour. In doing so, it takes in heat from the inside of the refrigerator. The vapour is then compressed by an electric pump, and heats up even more.

▼ **An outline of the way in** which a refrigerator works. The refrigerant goes through an endless cycle of evaporation (in the evaporator), compression (in the compressor) and finally condensation (in the condenser).

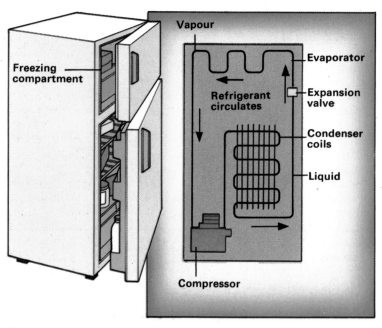

Vapour

Freezing compartment

Refrigerant circulates

Evaporator

Expansion valve

Condenser coils

Liquid

Compressor

► **Liquid nitrogen, at** a temperature of about −190°C, is often used to store biological materials. Here sperm is being stored. At such low temperatures it can be kept almost indefinitely without deteriorating.

The hot compressed vapour next passes through the coils of a condenser on the outside of the refrigerator. The coils give off the excess heat to the air. As the vapour cools down in the coils it condenses, or turns back into liquid. This circulates up to the expansion valve in the evaporator, and the cycle begins again.

A similar process of compression and expansion can be used to produce very low temperatures indeed. Even the air around us can be liquefied, or turned into liquid.

Air is liquefied by a repeated cycle of compression (at pressures 200 times as great as the normal atmospheric pressure), cooling and expansion. On each expansion, the air's temperature drops further until finally, at a temperature of about −190°C, it turns liquid.

From liquid air, liquid oxygen (for powering rockets) and liquid nitrogen (for storing biological specimens) can be obtained. So can several of the noble gases (see page 49), including argon (for filling light bulbs).

The science concerned with the production and application of very low temperatures is called cryogenics. In this field, scientists have discovered some very strange effects. Many metals, for example, lose all their electrical resistance and turn into 'superconductors' at extremely low temperatures.

Sound

If you hold your fingers against your throat when you speak, you will discover the secret of sound. It is made by vibrations – in this case the vibrations of your vocal chords. They pass on the vibrations to the molecules of the air, which carry them away.

When someone else speaks, vibrating molecules carry the sound of their voice into our ears. They make our eardrums vibrate, and we hear the sound of the speaker's voice.

The sound that reaches our ears may be pleasant, in the form of music, or unpleasant, in the form of noise. If the noise is too loud, it can cause physical pain when we listen to it.

The science of sound is called acoustics. Acoustic scientists and engineers are concerned with studying sound and systems for producing and reproducing it, with the design of concert halls, and so on. In the concert hall, for example, they must try to eliminate excessive echoes and other sound effects which might spoil the music.

Other acoustic engineers use echoes to advantage in designing sonar underwater communications devices for ships and submarines. They employ a kind of 'silent sound', beyond the range of human hearing, imitating a method bats use.

For music lovers, the faithfulness, or fidelity, of sound reproduction is everything, whether they listen on disc or tape. In recent years the compact disc has given us access to almost perfect high-fidelity ('hi-fi') reproduction. The sounds that emerge from the loudspeakers of a compact disc hi-fi unit are almost identical to those that entered the microphone in the concert hall.

▶ **A group of drummers** taking part in a street carnival in the South American city of Salvador de Bahia. Drums were one of the first musical instruments played by man, producing a basic rhythm to accompany dances and, later, other instruments.

Shaking the Air

Sound is set up in the air when an object vibrates, or moves back and forth, and passes on the vibrations to the air molecules. If there is no air in contact with a vibrating object, then the vibrations cannot be passed on. That is why there are no sounds in space, because space is an airless vacuum.

Sounds are not carried through the air just by the molecules which are in contact with a vibrating object. These molecules simply vibrate like the object, and pass the movements on to the molecules they are in contact with, which pass them on to the next ones, and so on. Each molecule in the 'chain' moves alternately forward and backwards in step with the vibrating

object. This causes a kind of ripple to spread through the air rather like a ripple spreads on a pond when you throw in a stone.

We call this sound ripple a sound wave. As we all know, there are many different sounds. That is because there are many different sound waves. As with all waves (see page 92), we can describe sound waves in terms of their wavelength and frequency (see box below).

Wavelength and frequency are related: a wave with a short wavelength has a high frequency; one with a long wavelength has a low frequency. We say that sounds with a high frequency, like those sung by a soprano, are high-pitched; those with a low frequency, like those sung by a

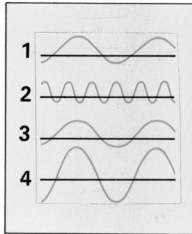

SOUND WAVES

Sound waves 1 and 2 have different wavelengths: the distances between their crests are different. Waves with long wavelengths take longer to pass a given point, so the waves have low frequency.

Sound waves 3 and 4 have the same wavelengths. However, they differ in their amplitude – the distance between their crests and troughs. The greater the amplitude, the louder the sound – see page 120.

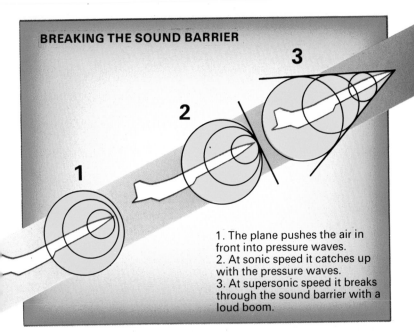

3

2

1

1. The plane pushes the air in front into pressure waves.
2. At sonic speed it catches up with the pressure waves.
3. At supersonic speed it breaks through the sound barrier with a loud boom.

bass, are low-pitched.

Mostly sounds are described by their frequency. It is measured in units called hertz (see page 93): 1 hertz (Hz) equals 1 vibration per second. A person with good hearing should be able to hear sounds between about 10 Hz and 20,000 Hz. We call sounds with a higher frequency ultrasonic (see page 123).

The speed of sound

In air at ordinary temperatures and at sea level, sound travels at a speed of about 1220 kilometres per hour. We call this sonic speed, or Mach 1. Sound travels faster in denser substances: over four times faster in water, and 17 times faster in steel.

When a jet plane flies through the air, it pushes the air in front of it to create pressure waves (picture 1 above). When it reaches sonic speed, it catches up with the pressure waves (2). When it goes supersonic – faster than Mach 1 – it leaves the pressure waves behind, setting up a shock wave (3). This travels through the air, creating a booming noise like a heavy gun firing.

Unless a plane is properly designed, the shock waves set up at Mach 1 can literally tear it apart. It was once thought that the speed of sound was the maximum speed limit for aircraft – this was a kind of 'sound barrier'. But carefully designed planes now reach Mach 3 or more.

Noise and Music

Wavelength and frequency are two characteristics of a sound wave. Amplitude is another. This is the amount the wave moves up and down (see the diagram on page 118).

The amount of energy in a vibrating object is reflected in the intensity of the sound it gives off. A big vibration packs a lot of energy, and will therefore create intense sound waves, with a large amplitude.

We describe the intensity of sound in terms of loudness, and measure it in units called decibels. On this scale, a whisper is about 20–30 decibels, and a supersonic jet taking off is about 120 decibels (see diagram below).

Any sound with a decibel level of over 140 is painful to the ears, which is why people in particularly noisy occupations (such as boilermaking) wear ear 'defenders' for protection.

Generally, we use the term 'noise' to refer to anything very loud. But in sound studies 'noise' has a rather different meaning. It means unwanted sounds, which may, for example, drown sounds we want to hear, damage our hearing, or generally interfere

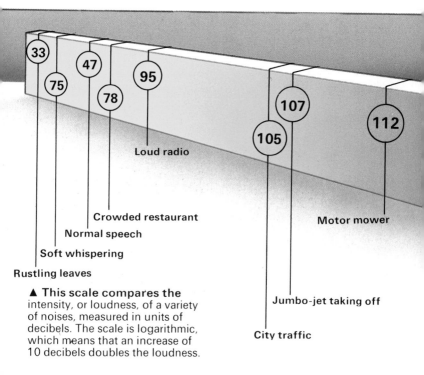

Loud radio

Crowded restaurant

Normal speech

Soft whispering

Rustling leaves

Motor mower

Jumbo-jet taking off

City traffic

▲ This scale compares the intensity, or loudness, of a variety of noises, measured in units of decibels. The scale is logarithmic, which means that an increase of 10 decibels doubles the loudness.

▲ **Cutting down trees with a** petrol-engined chain-saw is one of the noisiest jobs there is. To prevent damage to their hearing, workers wear ear defenders.

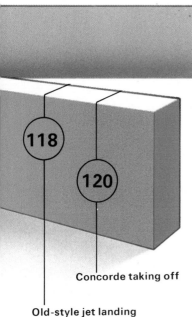

118

120

Concorde taking off

Old-style jet landing

with whatever we are doing. It is not necessarily the loudness of a noise that causes the problem. Usually it is the nature of the sound waves themselves – they are often irregular and jar the ear. The rattling of a loose window frame in the wind is an example of an object that creates irregular and unpleasant noise.

By contrast, sounds produced by regular vibrations are pleasant to the ear. We hear them as music. Musical sounds are produced by something vibrating: strings (violin and piano); a column of air (recorder and organ); a skin (drum); or a piece of metal (triangle).

121

Sound Effects

When a police car races past you with its siren blaring, notice how the sound changes. The sound you hear when the car is approaching has a higher pitch than that you hear when it is moving away.

The change in the pitch, or frequency, of the sound emitted by the siren is known as the Doppler effect. It is easily explained. When the siren is coming towards you, the sound waves are pushed together more. So more of them reach you in a given time. In other words, they appear to have a higher frequency. When the siren is going away, fewer waves reach your ears in a given time. In other words, they appear to have a lower frequency.

Sopranos with particularly pure and powerful voices have often been known to shatter wine

▼ **The ancient Greeks knew a** great deal about acoustics, the science of sound. Their amphitheatres were specially designed so that the audience could hear the actors clearly wherever they sat.

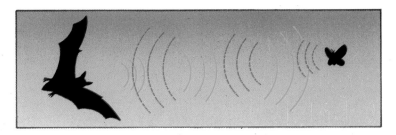

glasses when they hit the high notes. This happens because of an effect called resonance. Some substances, including glass, have a natural frequency at which they will vibrate when they are struck. When a sound wave reaches them vibrating at their natural frequency, they too will begin to vibrate in sympathy. This is resonance. If the energy in the wave is high enough, the substance can literally shake itself to pieces. This is what happens when a wine glass shatters.

Echoes

Just like light – another kind of wave – sound can be reflected. We hear reflected sounds as echoes. We often hear echoes in a

▼ **Ships hunt for submarines** by sonar – sending out sound waves and listening for echoes.

▲ **Bats emit high-pitched** squeaks to detect their prey. When they receive back echoes, they know that a meal is nearby.

large building such as a church, and in valleys and canyons outdoors. Designers of concert halls, for example, must be careful to avoid echo effects that will spoil the sound of the music.

On the other hand, echoes are used deliberately in the underwater communication and detection method called sonar. This uses ultrasonic waves – sound waves of frequencies higher than we can hear. Using this technique, it is possible for one submarine, for instance, to locate another by bouncing ultrasonic waves off it and listening for the echoes. This 'echo-location' is similar to the method bats use to navigate and locate their prey.

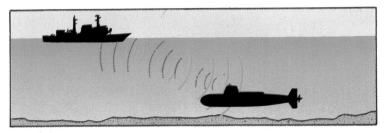

Reproducing Sound

Sound has a fleeting existence. But we can store sound waves on a tape or disc if we first convert them into a different form. On tape they are stored as a magnetic pattern. On an ordinary record disc they are stored in wavy grooves; on a compact disc, as microscopic pits (see pages 126–7).

These methods of storing sounds are quite different, but they all begin in the same way – with a microphone. And in each case, the sounds are retransmitted by a loudspeaker.

A microphone changes a pattern of sound waves into a matching electrical pattern in the form of a varying electric current. In a common type of microphone, a thin disc, or diaphragm, is attached to a piezoelectric crystal. This is a crystal which gives out a tiny electric current when squeezed. Sound waves entering the microphone cause the diaphragm to vibrate. This puts pressure on the crystal and sets up a tiny electric current. The current varies according to the vibrations, and holds an 'image'

► **(Top) This is a** crystal microphone. Sound vibrates the diaphragm, which in turn vibrates the crystal. This sets up electrical signals that represent the sound waves.

► **(Bottom)** A moving coil loudspeaker. When electrical signals pass through the coil, the cone vibrates and gives off sound waves.

◄ **In a recording** studio, sounds are picked up by several microphones. The microphone signals are adjusted on a mixer unit (foreground) to produce a balanced sound, which is then recorded on magnetic tape.

CRYSTAL MICROPHONE

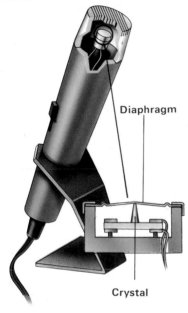

Diaphragm

Crystal

MOVING-COIL LOUDSPEAKER

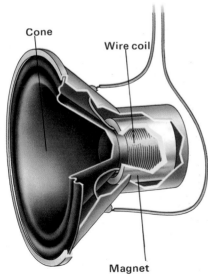

Cone

Wire coil

Magnet

of the sound waves. The 'image' is then stored on disc or tape.

At the listening end of the sound reproduction process, the 'image' is recreated as a varying electric current, similar to that which was produced by the microphone. This current then enters a loudspeaker. The standard loudspeaker consists of a paper cone with a coil of wire around its narrow end. This end is located inside a magnet.

The varying electric current representing the sounds is fed through the coil. This causes the coil to become magnetic (see page 140) and push and pull against the surrounding magnet. The cone vibrates and gives out sound waves similar to those that entered the microphone in the first place.

Record discs

The principles of sound recording on an ordinary record disc follow those worked out by Thomas A. Edison in 1877. This prolific American inventor made the first successful recording, using a so-called 'phonograph'.

To make an ordinary record disc, the varying electrical signals from the recording microphone are made to vibrate a needle, or stylus. This cuts a wavy groove in the lacquer surface of a spinning master disc. The waves in the grooves hold an 'image' of the sound waves that originally entered the microphone. The black plastic record discs we buy are

125

identical copies of this master.

To play back the record on a record player, the recording process is reversed. The stylus of the pick-up head of the player is placed in the groove in the disc, and the disc is spun. The waves on the sides of the grooves make the stylus vibrate. A device (often a crystal) in the head changes the vibrations into electrical signals. These go to a loudspeaker, which reproduces the sound.

▼ **On a record disc, sounds** are recorded as waves in the sides of the grooves. The needle of the record player vibrates as it travels along the grooves, and these vibrations are converted into electrical signals.

Tapes

In the case of a tape recorder, the electrical signals from the recording microphone are first converted into magnetic signals. This is done by a tiny electromagnet, called the record head. An electromagnet only becomes magnetic when electric current is passed through its coils (see page 141).

A tape coated with magnetic particles goes past the head and becomes magnetized in a pattern that represents the original sounds. On playback, the tape goes past another electromagnet called the playback head. The changing magnetic pattern on the tape sets up a varying electric

RECORD DISC

MAGNETIC TAPE

Close-up of grooves

Close-up of tape

After recording

Before recording

Record player

Cassette tape

current. This is identical to that which made the pattern in the first place. So when the current is fed to a loudspeaker, the original sounds are reproduced.

Compact discs

The electrical signals from the microphone are recorded on a compact disc by means of a laser beam (see page 106). A laser is used for playback too. The signals are recorded precisely on the disc in the form of numbers. This process is so accurate that sounds can be reproduced nearly perfectly. The compact disc is only about 12 cm across but it gives as much playing time as a 30-cm long-playing record.

In the recording process, the microphone signals are converted into numbers on the binary system – the one used in computers (see page 151). They become a series of ones and zeroes. The number signals are then fed to a laser, which cuts tiny pits in the surface of a master disc. The pattern of pits in the surface represents the sounds recorded very precisely.

In a compact-disc (CD) player, a laser scans the disc and is reflected. The reflected beam, holding the pit pattern, is passed through a 'photo-electric' device which converts the light signals into electrical ones. These are then fed to a loudspeaker.

◀ **In a tape recorder, sound is** recorded on magnetic tape. The magnetic particles on the tape are arranged in a pattern that represents the sound.

▶ **On a compact disc, sound is** recorded in the form of a spiral pattern of microscopic pits. In a CD player, a laser beam is used to read the pattern.

COMPACT DISC

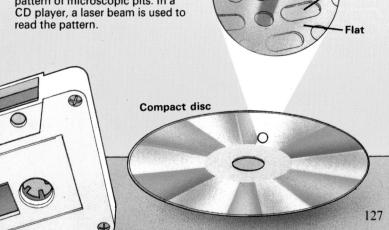

Laser beam

Close-up of disc

Pit

Flat

Compact disc

Electricity and Magnetism

Electricity is our most useful form of energy. It is available to us at the flick of a switch anywhere we can bring in a wire. We use it in light bulbs to lighten our darkness; in fires and cookers to produce heat; and in all kinds of applications to produce motion, from electric toothbrushes and vacuum cleaners to milk floats and locomotives.

The ancient Greeks first investigated the phenomenon we call electricity over 2500 years ago. They discovered that rubbing pieces of amber made it able to attract other things. This effect came to be called electric, after 'elektron', the Greek word for amber.

We call the kind of electricity shown by amber and similar substances static electricity. It is all to do with electrical charges on the atoms of the substances. They become charged when they gain or lose electrons.

In a substance charged with static electricity, the charge is 'trapped' in place. But electricity need not be stationary. It can be put to use by allowing it to flow through certain materials called conductors, where the electrons are free to move from atom to atom. The electricity produced by chemical reactions in a battery can be made to flow through metal wires to where it is needed – to light up a bulb, for example.

Current electricity is by far the most useful form. It can be made in batteries and generators. And it can be produced on a large scale at power stations and easily distributed to where it is needed.

Electric generators and the electric motors that power all our electrical appliances also rely upon the magnetic effects of an electric current. This is called electromagnetism. Some materials are magnetic even when no current flows. Their (static) magnetism can be seen as being parallel to static electricity. And magnetic effects are indeed similar to the effects of electric charges.

▶ **New York City at night:**
modern life would be impossible
without the aid of electricity.

Static Electricity

If you rub a blown-up balloon against your sweater and hold it near some bits of paper, they will stick to the balloon. Why is that? They stick to the balloon because it has become electric, or, rather, gained an electric charge. If you comb your hair vigorously, you make the comb electrically charged too. If you place it above your head, your hair will stand on end – not because you are frightened but because it has become charged as well. The electricity of the balloon and the comb is static electricity

The behaviour of the rubber balloon can be explained in terms of its atoms. Normally the atoms of rubber are electrically neutral – they have the same number of positively charged protons in the nucleus as negatively charged electrons circling around it (see page 26).

But rubbing the balloon on your sweater causes some electrons to move from atoms in the sweater to rubber atoms in the balloon. In other words, it causes the transfer of electric charges. The rubber ends up with extra negative charge, while the sweater ends up with extra positive charge.

When the balloon is placed near the bits of paper, the extra negative charges will cause positive charges in the paper to move towards them. It is a basic rule of static electricity that unlike (different) charges attract one another. And it is this attraction that causes the paper to stick to

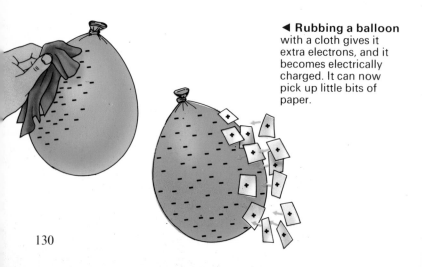

◀ **Rubbing a balloon** with a cloth gives it extra electrons, and it becomes electrically charged. It can now pick up little bits of paper.

the balloon. Another basic rule is that like (similar) charges repel one another.

▲ A spectacular display of lightning. The zig-zag paths show where high-voltage static electricity is streaking to the ground.

Lightning

Rubbing of a different kind goes on in thunderclouds but produces similar effects. Water and ice particles race about in the clouds, creating electric charges. These charges can build up to a tremendous 'pressure', or voltage (see page 133). To relieve this pressure, a lightning flash occurs. The flash marks the path of the electricity to the ground. The flash heats up the air and causes it to expand explosively. The pressure waves set up in the air hit our ears and we hear the familiar crack or rumble of thunder. We see the flash before we hear the thunder, because light travels much faster than sound.

▼ A lightning conductor has two effects. It causes positive ions in the air to move towards the cloud to help neutralize its charge. It also attracts electrons in lightning and channels the electricity safely to the ground.

Flowing Electricity

When you comb your hair with a plastic comb, you set up electric charges in it, which can make your hair stand on end (see page 130). Try combing your hair with a metal comb instead. What do you find? This time you cannot charge it up. This must mean that any electrons that might be moved by the rubbing action flow back again immediately into the metal.

Metals allow electrons to flow freely through them. They are a class of materials known as conductors. Carbon, in the form known as graphite (see page 47), is another notable conductor. But most other materials do not conduct electricity. We call them insulators – glass and rubber are examples. A third class of materials will conduct electricity a little when they are 'doped' with certain impurity atoms (that is, atoms of a different nature, which do conduct). They are called semiconductors and are vital to modern electronics (see page 146).

Metals are good conductors because they are made up of atoms whose outer electrons (see page 25) can wander. There is a 'sea' of electrons in metals, which, under suitable conditions, can move this way and that. It is the flow of electrons that makes up an electric current.

Just as there needs to be a

◄ **A worker in an** electricity substation (see page 145). Massive ceramic insulators are used to stop the dangerous current leaking out from the cables and causing damage or injury.

▶ **In this factory,** coffee pots are being plated with silver by means of electricity. They are placed in a silver solution. When electrical current is passed through the solution, pure silver is deposited on them.

MICHAEL FARADAY

The English chemist Michael Faraday (1791–1867) was one of the great pioneers of current electricity. He is perhaps best known as the inventor of the electric generator, the means by which continuous electric current can be produced. This was part of his investigations into electromagnetism (see page 141). He also studied the chemical effects of electricity.

difference in pressure to make a water current flow, so there has to be a difference in electrical pressure to make an electric current flow. We call this the potential difference, and we measure it in units called volts. The electrical pressure is provided either by a battery (see page 134) or else by a generator (page 144).

A battery produces electricity as a result of chemical reactions. The opposite effect can also take place. Electricity can be used to bring about reactions, especially the breakdown of substances. This process is called electrolysis.

Batteries

We can think of a battery as being a pump which supplies the pressure, or voltage, to keep an electric current flowing in a wire.

In 1800, an Italian scientist named Alessandro Volta made the first battery, which could provide a steady supply of electricity. It became known as the voltaic pile. It was made up of a pile of silver and zinc discs, with layers of pasteboard soaked in salt in between.

Among the early imitators of the pile was the simple cell. This used zinc and copper rods, dipping into a solution of sulphuric acid. We can outline the way batteries work by looking at the processes inside the simple cell.

In the solution, sulphuric acid

Silver disc

Salt-soaked pasteboard

Zinc disc

▲ **Volta's voltaic pile, which** consisted of alternate discs of silver and zinc, with salt-soaked pasteboard in between.

is present as ions (see page 51) – particles that carry electric charge. There are hydrogen ions with a positive electric charge, and sulphate ions with a negative electric charge.

ALESSANDRO VOLTA

Alessandro Volta (1745–1827) was the Italian inventor of the battery. He worked out the principles – of using two different metals and a conducting solution – when trying to explain an experiment with frogs' legs carried out by his fellow countryman Luigi Galvani. When Galvani hung the legs by brass hooks from iron railings, they twitched when they touched the railings. Galvani declared this was due to 'animal electricity' or galvanism. Volta believed that the electricity was produced when the two different metals were connected by the legs, whose moist flesh acted as an electrolyte. Alas for Galvani, Volta was right.

A few atoms from the zinc rod pass into the solution as ions with a positive charge. They leave a few electrons (negative charge) behind. When the zinc rod is connected to the copper rod by a wire, these electrons flow through it to the copper. There they attract the positively charged hydrogen ions. The hydrogen ions gain an electron and become hydrogen atoms (this process is explained on page 58), while hydrogen gas bubbles off.

Meanwhile, more zinc ions have entered the solution, leaving more electrons behind. These flow through the wire to the copper rod, and the cycle continues. The result is a continuous flow of electrons through the wire: in other words, an electric current.

We call the zinc rod, which is the source of the electrons, the negative electrode, or cathode. The copper rod is known as the positive electrode, or anode. Although the electrons flow from the negative to the positive electrode, we say by convention that the current flows from positive to negative.

All batteries work on similar principles, but use different electrodes and different electrolytes, or conducting solutions. The familiar dry cell used in torches, for example, uses carbon and zinc electrodes and a conducting paste for the electrolyte. A car battery uses lead and lead oxide electrodes, in a sulphuric acid electrolyte.

The dry cell and cells like it, called primary cells, stop working when their chemicals are used up. But the lead-acid car battery, known as a secondary cell, can be recharged. By putting in electricity, the electrodes and electrolyte can be changed back into their original condition, so they can again produce current.

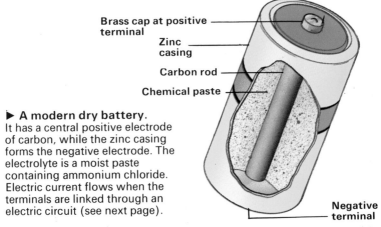

Brass cap at positive terminal

Zinc casing

Carbon rod

Chemical paste

Negative terminal

▶ A modern dry battery.
It has a central positive electrode of carbon, while the zinc casing forms the negative electrode. The electrolyte is a moist paste containing ammonium chloride. Electric current flows when the terminals are linked through an electric circuit (see next page).

Electric Circuits

In a battery, no changes take place until the two electrodes are connected by a wire. Then the electrons can flow from the negative to the positive electrode. The wire forms a simple pathway, or circuit, for the electrons. A circuit may also include other components (parts) as well. But as long as the electrons can pass through them, an electric current will flow in the circuit.

As electrons flow through conductors, such as metals, they tend to bump into the metal atoms. In doing so they lose some of their energy to the atoms. In other words they meet resistance, and the effect is called electrical resistance. It slows down the electron flow, reducing the electric current. At the same time the metal atoms gain energy and start to vibrate faster. This makes the metal hotter, because the hotness of a body is related to the speed of its atoms or molecules (see page 108). The amount of electrical resistance and heating varies according to the type and dimensions of the conductor.

The units of electrical 'pressure', current and resistance are named after electrical pioneers.

▼ **Light bulbs can be wired** into a circuit in two different ways. They can be wired in a series circuit (left). But if you take out one bulb, it breaks the circuit and the others go out. The bulbs can also be wired in parallel (right). If you take out one bulb, the others will still stay on.

SERIES CIRCUIT   PARALLEL CIRCUIT

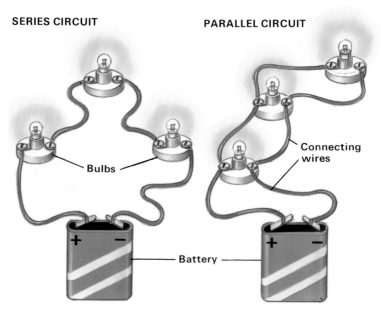

Bulbs

Connecting wires

Battery

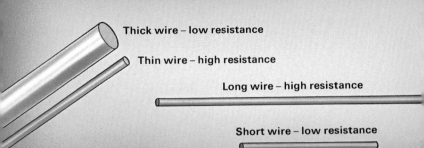

Thick wire – low resistance

Thin wire – high resistance

Long wire – high resistance

Short wire – low resistance

The unit of 'pressure' is the volt, named after Alessandro Volta (see page 134). The unit of current is the ampere, or amp, named after the French scientist André Marie Ampère. The unit of resistance is the ohm, named after the German Georg Simon Ohm. In 1827, Ohm discovered a relationship between voltage, resistance and current in a circuit: that resistance equals voltage divided by current. This is called Ohm's law.

Torches

A torch is a familiar example of a simple circuit. It consists of a battery linked by a metal conductor to a light bulb. The light bulb contains a wire with high electrical resistance, which heats up and glows when current flows through it. The circuit has another essential component, a switch. This is needed to break the circuit – to cut off the current to the bulb – when you are not using the torch.

▲ **Different sizes of wires** have a different electrical resistance, allowing a different flow of current. This is just like the different water flows allowed by different sized water pipes.

▼ **Many Christmas tree lights** are wired in a series circuit, so if one goes out, they all go out.

Electric Lighting

When electric current passes through a wire, heat is given out because of the wire's electrical resistance. If the current is strong enough, the heat given out will make the wire glow. This is the principle behind the electric light bulb.

Joseph Swan in Britain demonstrated the first light bulb in 1878. It was a glass bulb containing, not a wire, but a thin thread of carbon to produce light. Air had to be removed from the bulb to stop the carbon oxidizing, or burning away. Thomas Edison in the United States made a similar lamp a year later.

In the modern light bulb, electricity is passed through a thin wire, or filament, made out of tungsten. Tungsten is chosen because it is the metal with the highest melting point (3380°C). The glass bulb is filled with a mixture of argon and nitrogen gases. They do not react with tungsten, and thus help prolong its life.

Discharge lamps

At ordinary pressures, gases are poor conductors. But at very low pressures, they do conduct elec-

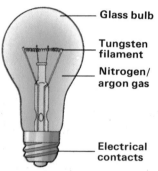

- Glass bulb
- Tungsten filament
- Nitrogen/ argon gas
- Electrical contacts

▲ **The fine wire filament of a** light bulb glows white-hot when electricity is passed through it.

▶ **In a fluorescent lamp,** mercury atoms give off ultraviolet light when their electrons jump from one orbit to another. This light makes the phosphor coating glow in visible light.

▶ **(Far right) Colourful** discharge lamps light up the casino city of Reno, USA.

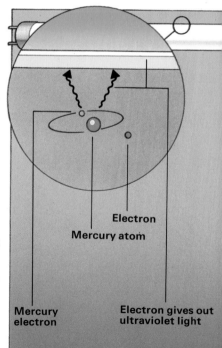

Electron

Mercury atom

Mercury electron

Electron gives out ultraviolet light

tricity in what is called an electric discharge. Michael Faraday (see page 133) was one of the first to investigate electric discharge in the 1830s. Today lamps using this effect are in widespread use. They include neon, sodium and fluorescent lamps.

Discharge lamps consist of a glass tube with electrodes sealed in each end to carry the electricity in and out. The tube is filled with gas at a pressure of only about one-thousandth of normal atmospheric pressure (see page 22).

When electric current is passed through the tube, electrons travel from one end to the other. As they do so, they collide with atoms of gas, and give them extra energy. Afterwards the 'excited' gas atoms give off this excess energy as light. The wavelength, or colour, of the light given off varies from gas to gas. Neon, the gas most widely used, gives off a vivid orange-red colour; sodium vapour glows a brilliant yellow.

Fluorescent lamps contain mercury vapour. This gives out invisible ultraviolet light in an electric discharge. So the discharge tube is coated inside with a layer of phosphors. They fluoresce, or turn the ultraviolet light into visible white light.

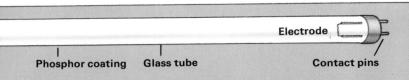

Phosphor coating   **Glass tube**   **Electrode**   **Contact pins**

Magnetism

If you spill a load of pins on a carpet, what is the easiest way to pick them up? With a magnet. A magnet attracts materials like steel, which is what the pins are made of. But when you try to pick up a milk-bottle top with a magnet you can't. That is because the top is made of aluminium, and aluminium is not affected by magnets. The only materials that can become magnets and be attracted by magnets are iron, cobalt, nickel and some alloys.

If you hang up a magnet so that it can swing freely, it always comes to rest pointing in a north–south direction. (Try it.) We call the end that points north, the north pole, and the end that points south, the south pole.

The magnetism of a magnet is concentrated at the poles. If you place one in a pile of iron filings (tiny bits of iron), the filings will stick mainly to the ends. They will also tend to arrange themselves in curves, as if following invisible lines. In effect, that is what they are doing – these are invisible lines of magnetic force.

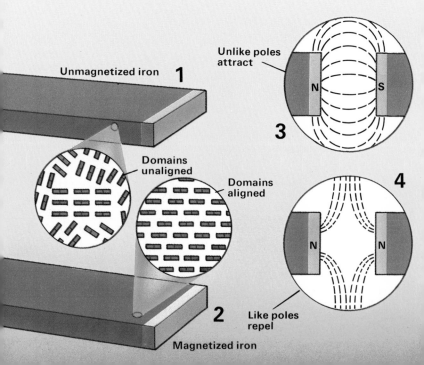

Unmagnetized iron **1**

Domains unaligned

Domains aligned

2 Magnetized iron

Unlike poles attract

3

N   S

4

N   N

Like poles repel

They form what is known as a magnetic field.

You can discover a basic law of magnetism yourself, using two magnets. First find out which are the north and south poles. Then bring together in turn two north poles, two south poles, and a north pole and a south pole. You will find that the two north and the two south poles will repel, or push away, each other. But a north and a south pole will pull together, or attract one another. We can now state a law of magnetism, that like poles repel and unlike poles attract.

The Earth magnet

A suspended magnet lines up north–south because the Earth itself behaves like a huge magnet, with one pole near the geographic North Pole and the other near the geographic South Pole. A magnet lines up with the invisible lines of the Earth's magnetic field. The needle of a compass is a weak magnet, which is why it always points north–south and can be used to find direction.

◄ **Iron is made up of areas** called domains, which contain tiny magnets lined up in one direction. In unmagnetized iron (1), the domains all point in different directions. But when it is magnetized (2), all the domains line up in the same direction. Unlike poles (3) attract each other, while like poles (4) repel. The lines of force are different in each case.

▲ **An electromagnet picking** up iron scrap. It consists of a wire coil wound around an iron core. It remains magnetic only while current passes through the coil.

In practice, a compass does not point exactly north–south because the magnetic poles are not exactly in the same positions as the geographic poles. The direction of the compass needle is actually about 12° out. This must be allowed for in map-reading.

The Earth's magnetic field extends far out in space, creating a kind of giant magnetic bubble called the magnetosphere.

Electricity and magnetism

Magnetism is closely related to electricity. When electricity flows in a wire, a magnetic field is set up around it. The study of this effect, called electromagnetism, has made possible such devices as electromagnets, electric generators and motors (see next page).

Motors and Generators

You can prove that an electric current passing through a wire sets up a magnetic field if you put a compass near a wire carrying a current. The compass needle will move when the current is switched on – it lines itself up with the magnetic field around the wire.

The first person to observe this effect was the Danish physicist Hans Christian Oersted, in 1820. And it was this observation that gave birth to the science of electromagnetism.

Consider what happens when a wire carrying an electric current (with its accompanying magnetic field) is placed next to a magnet. The magnet, of course, has its own magnetic field. So we have two magnetic fields close to each other. They must attract or repel one another according to the law of magnetism. The force of attraction or repulsion makes the wire move, if it is free to do so.

Summing up, we can say that when a current is passed through a wire in a magnetic field, the wire will tend to move. This is known as the motor effect, and is the principle behind the electric motor.

The principle of the electric motor is shown in the diagram below. This is a simple motor that works off the one-way, or direct current from a battery.

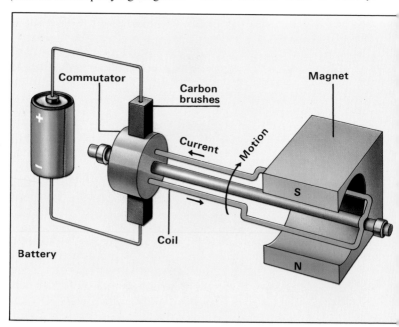

The interaction of the magnetic field of the coil and that of the magnet causes the coil to spin round. In a practical motor very many coils would be wound around the spindle, or armature. And the magnet might be an electromagnet, again made up of many coils.

The motor effect can be reversed. In other words, if you move a wire which is situated in a magnetic field, then an electric current will flow in it. This is called the dynamo effect, and is the principle behind the electric generator.

A generator is designed in much the same way as a motor, only the coils are driven round between the poles of the magnet.

▲ An electric motor powers this handy food mixer.

THE ELECTRIC MOTOR

In the battery-powered motor shown here, the wire is wound in a coil and positioned between the poles of a U-shaped magnet. The battery puts out one-way, or direct current (DC). The current magnetizes the coil, and it turns to align its north and south poles with the magnet. Inertia (see page 67) keeps it turning past the point of alignment. The commutator has now reversed its connection with the carbon brushes, so the direction of the current flowing through the coil is reversed. The coil is now repelled from the pole it has just passed, and attracted to the other – so it keeps spinning.

The diagram on the left (without the battery) can also represent a generator. The current will flow when the coil is turned. Because of the commutator, the current always flows out of the brushes in the same direction. That is, it is direct current (DC).

If a complete ring is used instead of a commutator, then the current coming from the brushes will reverse direction at each half-turn of the coil. It will produce alternating current (AC). The mains electricity that comes to our homes is AC. Motors working from AC do not need a commutator because the current changes direction at each half-turn of the coil, keeping the coil moving in the same direction.

Mains Electricity

The electricity that flows into our homes is called mains electricity. In most countries it is an alternating current, with a voltage of about 240 volts. It alternates (flows first one way, then the other) 60 times a second in North America, and 50 times a second in most parts of Europe. This is called the mains frequency.

Generating electricity

Mains electricity is generated at power stations and then transmitted overland by cables to the factories, businesses and homes that use it. At a power station, the electricity is produced by generating machines called turbo-alternators. They are turbines coupled to generators that produce alternating current (see page 143). The turbines are spun by steam in conventional and nuclear power stations, and by flowing water in hydroelectric power stations.

Typically a turbo-alternator

Mains electricity packs a lot of power. It must be handled with extreme care. At home you must never touch the insides of light or power sockets or bare wires when the current is switched on. An electric shock from the mains can kill you.

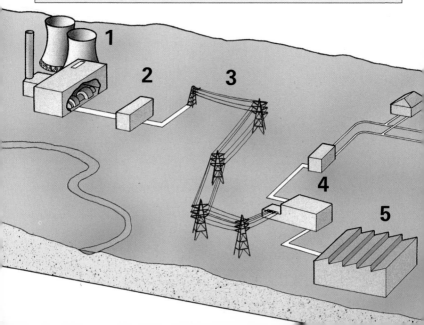

STEP UP, STEP DOWN

A transformer can be thought of as a kind of square doughnut, with coils of wire on two opposite sides. One coil has many more turns of wire than the other. When a low alternating voltage is fed to the coil with few turns, high voltage will be set up in the coil with many turns. This is the principle of the step-up transformer. On the other hand, when a high voltage is fed to the coil with many turns, a low voltage is set up in the coil with few turns. This is the principle of the step-down transformer.

produces alternating current at a voltage of 25,000 volts. It is not transmitted at this voltage because it would lose too much power in the transmission lines (cables). So first the voltage is boosted to 400,000 volts or more. This is done by a step-up transformer. Transmission lines on pylons carry the high-voltage electricity to substations, which use step-down transformers to reduce the voltage to the level needed by factories and homes.

To even out the electricity supply in a region, the power stations are connected in a distribution system called the grid.

◀ **Mains electricity is** produced at a power station (1). A step-up transformer (2) increases its voltage for transmission through lines (3) carried by pylons. The lines deliver it to substations (4), where step-down transformers reduce its voltage. The electricity is then fed to factories (5) or through another step-down transformer to ordinary homes (6). Usually a power station feeds electricity into a large distribution network called the grid. Power in the grid is switched to where it is needed at any time.

145

The Electronic World

Electrons are among the tiniest of the particles inside the atoms that make up matter. When they move, they set up an electric current. Electric current is the most useful form of energy there is. It flows through wires to heat up our cookers, turn on our lights, and power anything driven by an electric motor, from shavers to locomotives. All these uses involve the massive flow of electrons through metal conductors, especially copper wires.

Electrons can also flow, though with some difficulty, through other materials, called semiconductors. The advantage of such materials is that they can be used to manipulate and control very precisely the flow of electrons. And they have made possible the modern electronics industry, which has brought us such devices as the digital watch, the pocket calculator and the home computer.

The most widely used semiconductor material is silicon, which can be processed so that a small piece, or chip, can contain millions of electronic circuits. It is such chips that hold the key to the fantastic things computers can do, from being the brains of robots to calculating the path of space probes to distant planets.

The world of electronics is also concerned with the production and transmission of radio waves, upon which our long-distance ('tele-') communications depend. It is concerned with the manipulation of electrons in vacuum tubes, particularly the cathode-ray tube. This forms the screen in radar stations and television receivers.

▶ The cockpit of a Boeing 747 airliner, showing some of the flight and navigation instruments and controls. They rely heavily on electronics for their operation. Other major electronic systems on board an airliner are the automatic pilot, radar and radio communications systems.

The Chip Revolution

The electronics revolution that is now upon us began in 1948. Before that time, electronic devices, such as the radio, relied on valves, or vacuum tubes, to control and amplify (strengthen) the flow of electrons in circuits.

In 1948, however, came the invention of the transistor. This was the brainchild of the American scientists John Bardeen, Walter H. Brattain and William B. Shockley. The transistor used small crystals of semiconductor to control and amplify the flow of electrons. They soon replaced valves. Their great advantage was that they were very much smaller than valves and consumed much less power. This led to the pocket-sized transistor radio.

However, the transistors were still wired into the same kind of circuits as before. The next breakthrough came in the 1950s, when scientists began putting several transistors in a single crystal, together with the connecting circuits. This type of arrangement is called an integrated circuit.

Since then, more and more transistors and more and more circuits have been crammed into crystals every year. The most common form of crystal now is a thin wafer of silicon about 7 mm square. This is the silicon chip, or microchip. It carries hundreds of thousands of microscopic transistors as well as other electronic components.

Chipmaking

Making chips is a lengthy process. The basic principle is to build up layers of silicon which

◀ **Workers testing** the circuits in individual chips on a silicon slice. They need to use microscopes because the chips are so small. Chip manufacture and testing take place in clean rooms to ensure that they do not pick up any dust, which could ruin the circuits.

have different electrical properties. When the layers are completed, different parts become transistors, resistors (components which resist the flow of electric current), and so on, according to the arrangement of the layers.

The layers are given their special electrical properties by doping, or introducing impurities into the silicon. Different dopants, such as boron and phosphorus, impart different properties. So that the dopants reach only the desired areas, doping is done through a mask.

Chips are made several hundreds at a time on a silicon disc up to about 10 cm in diameter. The mask used for doping is first drawn about 250 times life-size and then photographically reduced. When processing is finished, the slice is cut up into separate chips. Each is mounted on a base fitted with pins to connect it to external circuits.

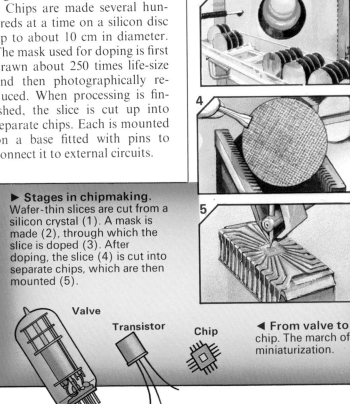

▶ **Stages in chipmaking.**
Wafer-thin slices are cut from a silicon crystal (1). A mask is made (2), through which the slice is doped (3). After doping, the slice (4) is cut into separate chips, which are then mounted (5).

Valve

Transistor

Chip

◀ **From valve to** chip. The march of miniaturization.

Calculators and Computers

The commonest use of the silicon chip is in the electronic calculator. This is in effect a miniature computer, and works in much the same way. All of its main electronic circuits are formed on a single chip. Some pocket-sized calculators are as powerful as a room-sized computer of the 1970s!

A home computer is much bigger than a calculator, but its heart is still just a single chip, called a microprocessor. Different parts of the chip carry out the different functions of a computer. These parts include two kinds of memory, a central processing unit (CPU) and a clock.

The read-only memory, or ROM, carries the information needed for the computer to work. It cannot be changed. The random-access memory (RAM) is a temporary memory, into which you feed instructions and

▼ A scientist operating the computer terminal of an instrument that analyses the structures of chemicals. Such work would not be possible without computers.

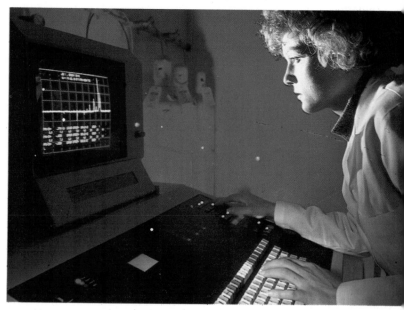

BINARY CODE

Binary code is a way of expressing words, symbols and numbers in digits in the binary number system. There are just two digits (0 and 1) rather than the usual ten (0 to 9) used in everyday counting. These two digits can be represented in a computer's circuits by, for example, the flow (1) or non-flow (0) of electric current.

128	64	32	16	8	4	2	1	
0	0	0	0	0	0	0	1	1
0	0	0	0	0	0	1	0	2
0	0	1	1	0	0	1	1	51
0	0	1	1	1	0	0	0	56
0	1	0	0	0	1	1	0	70
0	1	0	1					0
1								6
0								

```
01000001 A      01100011 c
01000010 B      01100100 d
01000011 C      01000101 e
01000100 D      00011110 f
01000101 E      00101011 +
                   101 -
```

information when you are operating the computer. The set of instructions you feed into a computer is called the program; and the information you put in is called the data. People who write computer programs use a simplified language so that the computer can understand. One of the commonest computer languages is BASIC.

The CPU controls the whole computer and directs the other parts to act according to the program. It includes an arithmetic logic unit, which carries out all the necessary calculations. The clock unit is a time base to ensure that operations are carried out in a regular order.

Computer software

The programs and data we feed into a computer are known as computer software. They can be fed in by a variety of means, including a keyboard. We can type in words and decimal figures, which go into the computer's RAM. But they do not go into the memory in the form of words and decimal figures, because the memory cannot store them in this form. The computer first needs to convert them into a binary code. This is a code which uses just two digits, 0 and 1 (see the box above).

The two binary digits are known as bits. Most computers handle and store data in the form of eight-bit units, called bytes. A typical home computer has a RAM that can hold up to about 128,000 bytes (or 128 kilobytes). The RAM of a desk-top office computer can hold several megabytes (millions of bytes).

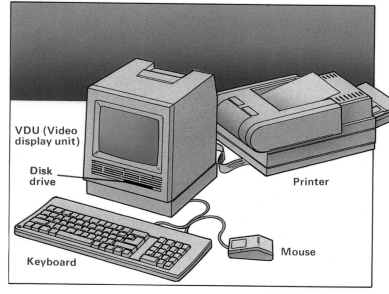

VDU (Video display unit)

Disk drive

Printer

Keyboard

Mouse

Computer hardware

The physical parts of a computer make up its hardware. In the home computer, this includes the main processing unit, a video display unit (VDU) like a TV screen, a keyboard, and maybe a mouse, joystick, printer and modem.

The processing unit has one or two slots in it to take magnetic disks carrying programs and data. They are called floppy disks because they are made of flexible plastic. The disks are driven round, rather like a record disc, when the computer is being fed information or when it is outputting results. Some computers have a built-in hard disk, which has a bigger memory.

The mouse is a device which you roll over a surface to direct

▲ **A typical microcomputer** set-up. Input is by keyboard, mouse or floppy disks in the disk drive. Words and graphics (pictures) are displayed on the VDU and can be printed out by the printer.

the cursor or arrow on the VDU. You use a joystick for much the same purpose in video games.

Results from computer operations may be outputted on the screen, printed out by a printer, or stored on floppy disk or magnetic tape. They can also be transmitted over the telephone lines to another computer. You do this with a modem, which changes computer output signals into audio (sound) signals to feed into the telephone mouthpiece. A modem at the other end converts them back again.

Robots

Robots, one of the favourite subjects of science fiction writers, are fast becoming scientific fact, thanks to computers. Robots can now be fitted with compact 'electronic brains', which can control them in a very complex way. They are now taking their place on the factory floor alongside human workers. There they pick up objects, assemble things, weld metal and spray car bodies.

The great advantage robots have over humans is that they can be put to work in conditions that humans would find intolerable or dangerous. Robots do not mind noise, glare, noxious fumes and dangerous radiation. Also they are tireless workers that do not get bored with endlessly repetitive jobs.

These robots are not like the androids of science fiction – they do not look like humans. They are built in whatever form is the most efficient to carry out their particular task. But they do have components equivalent to those in the human body. They have claws or grippers for 'hands', and swivel joints for 'wrists', 'elbows' and 'waist'. Their 'muscles' are powered by hydraulic (liquid-pressure) systems (see page 18), or else compressed air or electric motors.

Most of the robots on the factory floor are 'blind', which limits their use. The next generation of robots will be able to see with electronic eyes and tackle a much wider range of tasks than robots can now.

Electronic robots still have a long way to go to match the dexterity, senses and intelligence possessed by humans. But they are advancing rapidly thanks to the miracle of the computer and its manipulation of electrons.

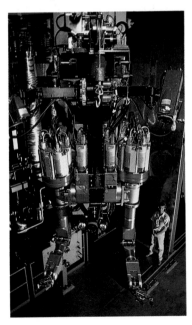

▶ **This advanced industrial** robot is used in the nuclear facility at Oak Ridge, USA. It has video-camera vision and is operated by remote control.

Telecommunications

Until the 1830s, the fastest way of sending messages a long distance was on horseback. Then William Cooke and Charles Wheatstone in England, and Samuel Morse in the United States, developed the electric telegraph. This sent messages as coded electrical signals through wires. In 1876, Alexander Graham Bell in the United States invented the telephone, which allowed voices to be carried along the wires. In 1894, Guglielmo Marconi in Italy invented the wireless telegraph to send messages long distances without wires. These people laid the foundations for today's telecommunications networks, which can flash messages around the world in a split second.

From the electric telegraph, the modern Telex system developed. From the wireless telegraph came modern radio. The telephone network too has developed so that it can now carry pictures, sent by facsimile (fax) machines, computer data and electronic mail. All these means of telecommunications rely heavily on electronics.

Radio broadcasting is not the only communications medium to

▼ **Air-traffic controllers at** their radar screens. Radar transmitters send out radio pulses, which are reflected back by aircraft. The echoes make luminous 'blips' appear on the controllers' screens.

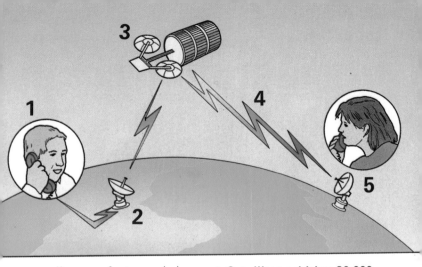

▲ **Satellites orbiting 36,000** km high now carry most overseas telephone calls. Signals from a caller's telephone (1) are beamed from a dish aerial (2) up to a communications satellite (3), which relays them to another dish (4) and on to the receiver (5).

use radio waves for transmission. Telex, telephone and television signals, and so on, may also travel via radio waves for at least part of their journey. For overseas communications especially, they may also travel via communications satellites in space. Usually very short radio waves, or microwaves, are used for such transmissions.

But just how can radio waves carry information? Let us take ordinary sound broadcasting as an example. The process begins when a person speaks into a microphone (see page 124). The microphone converts the sound waves into varying electrical current, or an electrical wave. This audio (voice) wave cannot be transmitted through the air because it is the wrong kind of wave.

Radio waves, however, can travel through the air. And they can also be made to carry the audio signals. The process of combining the two is called modulation. The radio carrier wave may have either its amplitude (see page 118) altered (amplitude modulation, or AM) or its frequency altered (frequency modulation, or FM) by the audio signals.

The modulated carrier wave is transmitted, and picked up by the aerial of the radio receiver. Electronic circuits in the receiver separate the audio signals from the carrier wave, a process called detection. When they are fed to a loudspeaker (see page 125), the sounds that originally entered the microphone are reproduced.

Television

Experiments that led to the development of television were being carried out over a century ago. The first working system was demonstrated in Britain by John Logie Baird in 1926. But his mechanical picture scanning system was soon replaced by an electronic one, developed by an American, Vladimir Zworykin.

Scanning is the process by which a television camera converts the scene it is viewing into a set of electrical signals. In the camera, light coming through the lens forms an image on a plate, which becomes electrically charged according to the pattern of light falling on it.

A beam of electrons from a so-called 'electron gun' is fired at the image, scanning from left to right and from top to bottom in several hundred lines. The result is a varying electric current, whose strength depends on the charge on the various parts of the plate, and so represents the light pattern in the scene viewed. The varying current that results

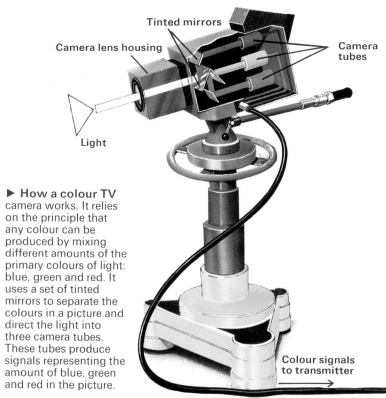

Tinted mirrors

Camera lens housing

Camera tubes

Light

▶ **How a colour TV** camera works. It relies on the principle that any colour can be produced by mixing different amounts of the primary colours of light: blue, green and red. It uses a set of tinted mirrors to separate the colours in a picture and direct the light into three camera tubes. These tubes produce signals representing the amount of blue, green and red in the picture.

Colour signals to transmitter

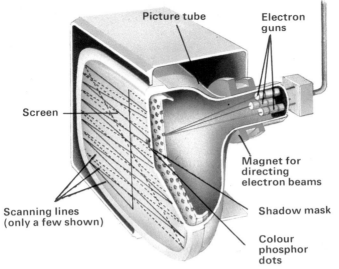

Picture tube

Electron guns

Screen

Magnet for directing electron beams

Shadow mask

Scanning lines (only a few shown)

Colour phosphor dots

▲ How a colour TV picture tube works. The three sets of colour signals representing blue, green and red are fed to electron guns, which fire beams of electrons at the screen. They pass through a shadow mask and cause the colour phosphors on the screen to glow.

forms the video (picture) signals, which are then transmitted on a carrier wave (see page 155).

The carrier wave is picked up by a TV aerial, and passes to the receiver. Electronic circuits separate out the video (picture) signals and feed them to the picture tube, which is a kind of cathode-ray tube. In the tube an electron gun fires a stream of electrons (known as cathode rays) at a fluorescent screen, making it light up when they hit. The beam scans back and forth in the same way as the beam did in the camera tube.

The video signals are made to increase or reduce the strength of the beam, making the screen brighter or dimmer. In this way a pattern of light is produced, which is similar to that viewed by the camera.

Colour TV

In a typical TV system, this process is repeated three times in order to transmit colour. Colour TV relies on the principle that any colour can be produced by mixing light of the three primary colours red, green and blue (see page 104).

In the camera, three tubes produce red, green and blue colour signals. The colours are recreated in the receiver's picture tube when three electron guns fire beams at phosphors that glow red, green and blue. From a distance, the colours merge to create a full-colour picture.

157

The Science of Life

The world is made up of things that are alive and things that are not. A poppy is alive; a pebble is not. When a poppy seed is sown in the ground, a plant grows up, flowers and produces more seeds. In other words, the poppy is able to reproduce itself. This is a characteristic of all living things. A pebble or any other non-living thing cannot do this. The science of living things is biology. Traditionally, it is split into two main branches: botany, the study of plants; and zoology, the study of animals.

All living things – plants and animals alike – are made up of basic building blocks called cells. Most plants and animals are made up of very many cells, of different types, which group together to form specialist tissues and organs.

Among these operations are making or processing food, to provide material for growth and to produce energy. Living things need energy to power the processes that sustain life and growth. Animals in addition need energy for movement. Respiration, or breathing, is an essential part of the energy-producing process.

Living things are involved in a constant fight for survival – in particular struggling to find enough food and water, and trying to avoid being killed by other living things. The more they can find out about the environment, or surroundings, in which they live, the better chance they have of surviving. Animals find out about their surroundings by using their senses.

All living things ensure the survival of their own kind, or species, by reproducing. Those best suited to survive pass on, through their genes, their own characteristics to their offspring. The study of genes, or genetics, has become one of the most exciting branches of science.

▶ **This awesome-looking** creature is not from another world. It is a Black fly from our own, pictured by a scanning electron microscope. The colours are false. Note all the separate lenses in its two large eyes.

Cells

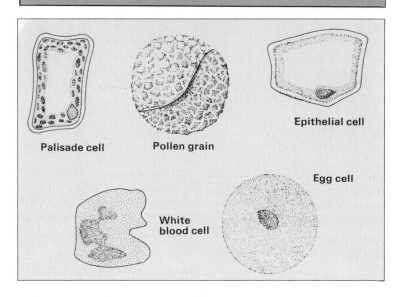

Palisade cell

Pollen grain

Epithelial cell

Egg cell

White blood cell

Cells are the basic body units from which all living things, or organisms, are made up. The simplest animals (such as the amoeba) and plants (such as algae) consist of just a single cell. But most living things are made up of very many cells. In multi-celled organisms different groups of cells combine together to form different tissues, such as skin and bone. In turn, different tissues combine to form organs (such as kidneys), which carry out certain jobs in the body of the organism (such as removing wastes).

All cells, however, are basically similar in both plants and animals. They consist largely of a jelly-like material called cyto-

▲ **A selection of plant (top)** and animal cells. In animals the white blood cells play a vital role in protecting the body against diseases.

plasm, held within a skin, or membrane. The most important part of the cell is the nucleus. This is the cell's control centre. It issues instructions to the other parts of the cell, such as the mitochrondria, which produce the cell's energy.

Plant cells differ from animal cells in having a rigid cell wall, made of cellulose. The cells in the green parts of the plant contain units called chloroplasts, which enable plants to make their own food (see next page).

Energy and Food

All living things must consume food to stay alive. Food provides the energy to make their bodies work, grow and, in animals, move. Plants are able to make their own food, a kind of sugar. Animals cannot. So, to survive, animals must eat plants or other animals that eat plants.

Plants make their food by a process called photosynthesis. They use simple ingredients: carbon dioxide, which they take in from the air (see page 164); and water, which they take in from the ground through their roots.

However, plants can only make food in sunlight, which provides the energy for the process. This energy is trapped by a substance called chlorophyll, which acts as a catalyst (see page 57) in the food-making process. Chlorophyll is a green pigment found in the chloroplasts of plant cells.

In the chemical reaction set off by chlorophyll, carbon dioxide combines with water to make sugar, while oxygen is given off

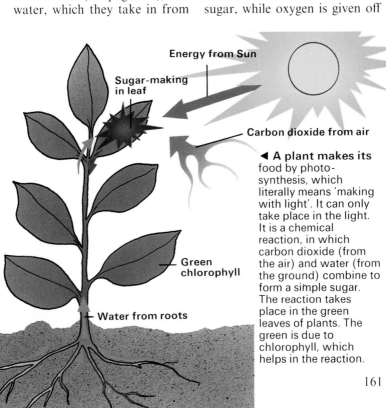

Energy from Sun

Sugar-making in leaf

Carbon dioxide from air

Green chlorophyll

Water from roots

◄ A plant makes its food by photo-synthesis, which literally means 'making with light'. It can only take place in the light. It is a chemical reaction, in which carbon dioxide (from the air) and water (from the ground) combine to form a simple sugar. The reaction takes place in the green leaves of plants. The green is due to chlorophyll, which helps in the reaction.

161

as a by-product. Plants use up some of the oxygen to 'burn' the food they have made to provide energy (see page 64). The rest is given off into the atmosphere.

Thus by taking in carbon dioxide and giving out oxygen, plants play a vital role in maintaining the balance of gases in the atmosphere. If too much vegetation is cut down over the world, as is currently happening in the Amazon rainforests, then this balance could be upset and all living things could suffer.

Feeding the animals
The bodies of animals are much more complicated than those of plants, and require a lot more energy to make them work. Animals also need a great deal of energy for movement. They thus

▲ A cheetah with prey in its mouth after a successful kill on the Serengeti Plains of East Africa. Cheetahs are the swiftest members of the big cat carnivores.

need to consume a lot of food. Many also need to eat different kinds of foods in order to take in the nourishing substances, or nutrients, that they need. The main nutrients are carbohydrates (such as sugar), proteins (found in foods such as meat and fish) and fats. Animals also usually need to eat traces of minerals and vitamins to remain healthy.

Simple animals do not actively look for food. Barnacles, for example, remain fixed to a rock, say, and filter specks of food from the water passing by. Advanced animals such as mammals do actively seek food, and often use up a lot of energy in doing so.

As far as feeding is concerned the animals fall into two groups – plant-eaters and meat-eaters. Cattle are typical plant-eaters, or herbivores. Lions are typical meat-eaters, or carnivores. The plant material that herbivores eat consists mostly of cellulose (see page 160). It is hard to digest (break down), and does not contain much energy. So in general herbivores have to eat often to get enough energy. Meat provides a lot more energy, so in general carnivores do not need to eat as often as herbivores.

As soon as food has been taken into the body, digestive processes begin to turn it into substances the body can use. In mammals, for example, digestion begins in the mouth and continues in the stomach and intestines. It is brought about by the chemical action of enzymes and acid. In the intestines, food substances pass into the blood, which carries them to all the cells in the body. Grazing animals such as cattle have a complex stomach with several chambers to digest the cellulose in grass. The first chamber is the rumen, and they are called ruminants.

Solid matter in the digestive system that the body cannot use is excreted as waste, or faeces. Liquid waste products pass out of the body as a solution in water, called urine.

SOURCES OF NUTRIENTS

▲ **Carbohydrates in** cereal crops and sugar provide energy.

▲ **Proteins in fish, meat,** beans and eggs help build body tissues.

▲ **Fats in meat, dairy** products and oils also provide energy.

▲ **Vitamins in vegetables** and fruit are needed to keep the body healthy.

Breathing

The energy living things require is produced when food, usually sugar, is 'burned' in the body cells. For this 'burning', or oxidation, process to occur, oxygen is required. The energy-making process also produces carbon dioxide as a waste product.

For an organism to remain alive, oxygen must be continuously supplied to the body, and carbon dioxide must be removed from the body. This process is called respiration, or breathing.

In plants, respiration generally takes place directly in the air through holes in the leaves, called stomata. In animals, which require so much more energy, respiration is usually more complicated. A number of methods may be used, according to the type of animal and the environment in which it lives.

In the air

All advanced animals have lungs, like human beings. Oxygen is taken into the lungs when the animal breathes in. Inside the lungs, oxygen passes into the blood, which carries it to the body cells. At the same time, carbon dioxide waste passes from the blood into the breathed-in air, and leaves the body when the animal breathes out.

In the case of most animals,

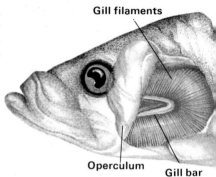

Gill filaments

Operculum   Gill bar

▲ **Fish breathe through their** feathery gills, which are located beneath a muscular flap called the operculum.

▼ **Plants breathe through tiny** holes in the leaves, called stomata. A stoma opens when the walls of the guard cell separate.

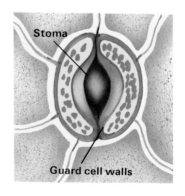

Stoma

Guard cell walls

enters the body through tiny holes called spiracles.

In the water

Fish use gills for breathing. Gills are feathery structures of fine blood vessels. Fish breathe by taking water through the mouth and expelling it past the gills. This allows oxygen dissolved in the water to pass into the blood and carbon dioxide to pass out. In most fish, the gills are internal, concealed beneath gill covers.

Some of the simplest water-dwellers, such as flatworms, have no respiratory system. Their flat body provides a large surface for oxygen to pass directly from the water into the body cells.

the movement of air into and out of the lungs is caused by changing their volume. This is usually done by altering the position of the rib cage and, in mammals, by contracting and relaxing the muscular diaphragm.

Birds are particularly active creatures, and they have a highly efficient lung system to ensure that they gain the maximum amount of oxygen each time they breathe in. Their lungs are connected to a number of chambers called air sacs. When they breathe, the air rushes through the lungs into the sacs and is completely changed.

Insects have no lungs. Instead, they have a system of fine, branching tubes, which carry air directly to the cells. The air

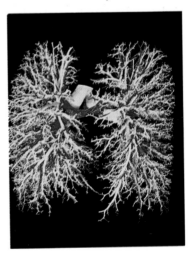

► **A model of the human** lungs, cast in resin. The airways branch into finer and finer tubes. From the air in the finest tubes oxygen passes into the blood; and into the air in the tubes carbon dioxide passes from the blood.

Senses

In order to survive, animals must be aware of what is happening around them. For example, they must be able to find food. They must also be able to detect the presence of predators that might harm them, so that they can hide or escape. And they must be able to find a mate so that they can have offspring to continue their kind.

Animals find out what they need to know about the outside world by using their senses. Many of them have the same main senses that we human beings have. They are sight, hearing, touch, smell and taste. The five senses are not all equally important to animals. For a sparrowhawk, for example, sight is most important, enabling it to detect the movement of small mammals in the grass while it is hovering maybe 50 metres up in the air. For a kiwi, which forages for food at night in the rain forests of New Zealand, smell is most important, enabling it to find its food even in the pitch darkness.

Many animals have the same kind of sense organs as humans – for example, eyes for sight and ears for hearing. Their eyes have sets of special 'receptor' cells which are sensitive to light. Information about the pattern of light falling on these cells goes to the brain, which creates a picture of the outside world. The ears contain a drum-like membrane, which vibrates when sound waves reach it through the air. Information on these vibrations

▶ **A Long-eared bat homes in** on its prey, an Underwing moth. It locates its prey by sending out high-pitched sound pulses and listening for any echoes.

◀ **A rattlesnake on the prairie** of Wyoming, USA. It detects warm-blooded prey by means of heat-sensitive pits on its head, just in front of and below the eyes.

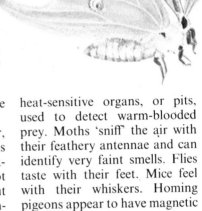

Antennae

▶ **This male Moon moth uses** its feathery antennae to detect the scent which is given off by female Moon moths.

is sent to the brain, and the animal hears.

Some creatures, however, have quite different sense organs from our own. Insects, for example, have eyes made up of not one lens, as in mammals, but thousands. They are called compound eyes (see page 159). Snakes 'smell', not with a nose, but with their tongue. Some have heat-sensitive organs, or pits, used to detect warm-blooded prey. Moths 'sniff' the air with their feathery antennae and can identify very faint smells. Flies taste with their feet. Mice feel with their whiskers. Homing pigeons appear to have magnetic sense organs, which enable them to navigate by detecting the Earth's magnetic field.

Reproduction in Plants

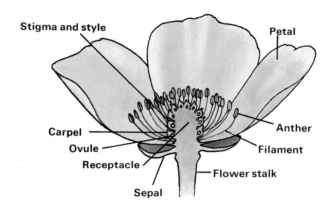

Stigma and style

Petal

Carpel

Anther

Ovule

Filament

Receptacle

Flower stalk

Sepal

By far the majority of plants reproduce sexually, as most animals do. They produce male and female sex cells. During the reproduction process the male cells fertilize the female cells. And 'offspring' are produced in the form of seeds.

In most plants the sex organs are contained in the flowers. The male parts are the stamens. They carry grains of pollen, which contain the male cells. The female part, the ovary, contains the female cells, the ovules. For fertilization to take place, the pollen must be transferred to the ovary. This process is called pollination.

Pollination and fertilization

Pollination may be brought about by the wind or with the help of insects, particularly bees. They pick up the pollen from the

▲ **The structures in a flower** that take part in reproduction. The anther is the male part that produces pollen, and with the filament makes up the stamen. Fertilization takes place when pollen from another plant reaches the female stigma and sends pollen tubes down to the ovules in the ovary.

stamens of one flower and fly to another. Some of the pollen rubs off against the stigma on top of the style, which extends from the ovary. When a pollen grain settles on the stigma, it sends down a tube to the ovary, carrying the male cell. This joins up with an ovule, bringing about fertilization. The ovule turns into an embryo, which develops into a seed.

The seeds of a plant may grow tiny 'wings', as in the case of the

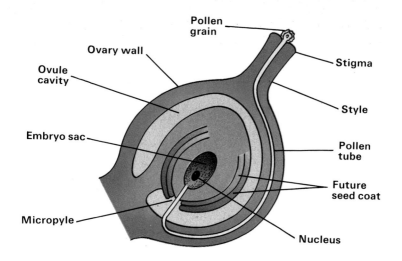

Pollen grain

Ovary wall

Ovule cavity

Embryo sac

Micropyle

Stigma

Style

Pollen tube

Future seed coat

Nucleus

▲ **A simplified diagram of a** single carpel in an ovary after fertilization. A pollen tube grows down from the stigma and enters the ovule through the micropyle. The tip of the tube breaks open to release two male nuclei. One fuses with the female nucleus to become the first cell of the new seed.

sycamore tree, or little 'para-chutes' as with the dandelion. Or the ovary itself may develop into a fruit, in which the seeds are surrounded by juicy flesh.

The reason for growing wings or parachutes, or developing into a fleshy fruit, is to bring about the better scattering, or disper-sion, of the seeds. A plant species stands a better chance of survival if its seeds are scattered over a wide area rather than if they simply drop beneath the parent

plant. With wings or parachutes, seeds can glide or be carried away on the wind. Fruits can be eaten by animals, which will carry the seeds some distance before they pass out on to the ground in the animals' droppings.

Asexual reproduction

Some flowering plants, however, can reproduce asexually, or with-out sex. The strawberry plant, for example, sends out long stems called runners. They take root, and new plants grow up. The tulip grows from a bulb and each year produces several new bulblets from which new plants will grow. The potato plant grows from a tuber, a swollen stem. It produces a large number of new tubers in the course of the growing season – these are the potatoes which we eat.

Egg

Larva (caterpillar)

Reproduction in Animals

Some of the simplest animals, such as the minute amoeba and hydra, reproduce asexually. The amoeba reproduces simply by splitting into two, a process called binary fission. The hydra reproduces by budding. Tiny 'buds' grow on its body, which eventually detach themselves and grow into new individuals.

Most animals, however, reproduce sexually. A male and a female get together so that their sex cells can join. The male cells are called sperm, and female cells are called eggs.

With some animals, such as amphibians and fish, fertilization takes place externally. The female first lays its eggs, and then the male releases its sperm over them. The process is often called spawning.

With amphibians, such as frogs, the fertilized eggs do not develop directly into young that are a replica of their parents. They develop first into a tadpole, which swims and breathes in the water. Then the tadpole changes into a miniature, air-breathing frog. Such a change of form during development is called metamorphosis.

In insects, reptiles and birds, fertilization is internal. The male mates with the female and introduces sperm into its body. And the eggs are laid already fertilized.

Insects, however, undergo a metamorphosis from egg to adult. With a butterfly, for example, the egg hatches first into a larva, or caterpillar. This then turns into a pupa, in which the adult butterfly forms. It emerges from the pupa and flies away.

Pupa

Adult

Mammals, including human beings, do not lay eggs. The embryo grows from the fertilized cell inside the female's body. It emerges – that is, it is born – when it is fully developed. The period of gestation, the time it takes between fertilization and birth, varies widely – from just 15 days for hamsters up to 22 months for elephants.

New-born mammals cannot eat adult food. They must feed on milk produced by the mother.

▲ **Insects such as this** Swallowtail butterfly undergo a change of form, called metamorphosis, as they grow. The butterfly starts life as an egg, which hatches into a larva, or caterpillar. In time the larva turns into a pupa. From the pupa the adult butterfly emerges.

▼ **An antelope licks its newly** born offspring. Unlike a human baby, the young antelope is able to run about almost immediately. This helps it survive in the wild.

Genetics

When a male and female organism – plant or animal – reproduce sexually, they pass on to their offspring certain of their own characteristics. For example, our parents pass on to us characteristics such as hair colour, height and shape of nose. We say we 'inherit' these characteristics from our parents.

The study of inheritance – the way characteristics are handed down through generations – is called genetics. Parents pass on characteristics by means of thread-like bodies called chromosomes, which are present in the nucleus of each cell. These bodies carry thousands of units called genes, each of which is responsible for a particular characteristic.

Each living species has a certain number of chromosomes in its cells. Human beings, for example, have 46 chromosomes in each ordinary cell, arranged in pairs. But in the sex cells – the sperm and egg – there are only 23 single chromosomes. When a

THE DOUBLE HELIX

The genetic information inside a chromosome is carried by a long molecule called DNA (deoxyribonucleic acid).

DNA is made up of two chains of alternating sugar and phosphate groups coiled around each other. They form what is termed a double helix. The two coils are linked by bonds between chemicals called bases, which are joined to the sugar groups. Each gene occupies a part of the DNA coil and consists of a unique sequence of bases.

◀ **The two scientists** who first worked out the structure of DNA in the early 1950s, James Watson and Francis Crick. For their work, one of the most important discoveries ever made in biology, they were awarded a Nobel Prize.

sperm fertilizes an egg, the two sets of chromosomes join together, and the resulting 'germ cell' has a full set of 46. These chromosomes carry genes from both the parents.

One pair of chromosomes inside the nucleus, called the sex chromosomes, determines whether an organism is male or female. In many living things, including humans, the males have two different sex chromosomes, called X and Y. The sex chromosomes of females on the other hand are both X.

The sex of any offspring depends on whether the sex chromosome in the fertilizing

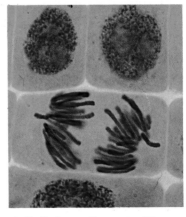

▲ **Cells in garlic roots. The** dark parts are the nuclei. The cell in the middle is undergoing mitosis, or division. Its chromosomes have just separated.

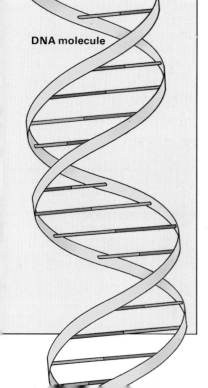

DNA molecule

sperm is an X or a Y. If it is an X, it will produce an XX in the offspring – a female. If it is a Y, it will produce an XY in the offspring – a male.

In recent years, genetic scientists ('geneticists') have been able to map individual genes on the chromosomes. And they have also developed techniques for removing and replacing genes. These techniques, called genetic engineering, have enabled them, for example, to produce improved species of crops, and valuable new medical products. The manufacturing of such products is known as biotechnology. In the future, geneticists expect that they will be able to eliminate many inherited genetic defects by removing or modifying the genes that cause them.

173

Glossary

Acid
A substance that provides hydrogen ions when dissolved in water. Acids combine with bases to form a salt and water only.

Acid rain
Rainwater which is more acidic than usual due to contamination from burning fossil fuels. It attacks buildings and can kill trees, plants and animals.

Acoustics
The science of sound. Also the sound properties of, say, a building.

Aerodynamics
The study of the behaviour of bodies moving through the air or other gases.

Alkali
A base that dissolves in water. The alkali metals are those whose hydroxides are strong alkalis. Sodium and potassium are examples.

Alloy
A mixture of a metal with another metal or with a non-metal. Brass is an alloy of copper and zinc; steel is an alloy of iron with small traces of the non-metal carbon, as well as other metals.

Atmosphere
The layer of air around the Earth, or more generally the layer of gases around any heavenly body. The atmosphere is also sometimes used as a unit of pressure, being the average

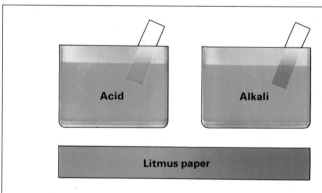

Indicators such as litmus paper are used to identify acids and alkalis. Blue litmus paper turns red in an acid solution. It remains blue in an alkali.

pressure at sea level exerted by the Earth's atmosphere.

Atom
The basic unit from which all matter is made up. It is the smallest part of an element that retains the properties of that element and that can take part in a chemical reaction.

Base
A substance that can accept hydrogen ions from an acid. A base reacts with an acid to form a salt and water only.

Capillarity
A property of liquids that makes them tend to rise or fall in narrow tubes. Water rises in glass tubes; mercury falls. Capillarity is an effect of surface tension.

Catalyst
A chemical substance that changes the rate of a chemical reaction, but is not itself chemically changed. Platinum, nickel and iron are all good catalysts.

Chemical bonding
The way in which atoms link together to form chemical compounds. Bonds are usually formed by means of the outer electrons of the combining atoms. For example, in ionic bonding, electrons are transferred from one atom to the other. In covalent bonding, each atom shares some of the electrons of the other.

Chemical compound
A substance made up of two or more elements chemically combined together. In the same compound the elements are always combined together in fixed proportions. For example, in the compound hydrogen oxide (better known as water), the elements hydrogen and oxygen always combine in the proportions 2:1.

Chemical reaction
A process in which substances undergo chemical change and form different substances. See also Catalyst.

Composition
The atoms or elements of which a substance is made up. Chemical substances have a fixed composition.

Compound
See Chemical compound.

Compression
A state in which a substance is subjected to pressure, or squeezing forces.

Condensation
The change of state that occurs when a vapour turns back into a liquid. When water vapour condenses into water droplets in the air, clouds form.

Conduction
The transmission of heat (thermal conduction) or electricity through a substance.

Convection

A means by which heat travels through a fluid – a gas or a liquid. The fluid expands as it is heated, and becomes lighter than the surrounding fluid. It therefore rises, starting an upward convection current.

Corrosion .

The breakdown of metals or other materials by chemical action. The rusting of iron and steel is the commonest form of corrosion, brought about by the combined action of the oxygen and moisture in the air.

Crystal

A substance that has formed naturally into a regular shape. Minerals form the most beautiful crystals when they solidify in cavities in the rocks. The regular external shape of a crystal relates to the regularity of its internal structure, called the lattice.

Decomposition

A chemical reaction in which a compound splits up into its elements or simpler compounds.

Density

The mass of a unit volume of a substance, such as a cubic metre. The most dense substance on Earth, the metal osmium, has a density of 22.5 kilograms per cubic metre.

The relative density of a substance is its density compared with that of water.

Diffraction

The spreading out of waves (including water, sound and light) when they pass through a narrow slit or around an obstacle.

Digestion

The process by which living things extract nutrients, or nourishing substances, from the food they eat.

Displacement

When a body is immersed in water, the displacement is the amount of water it pushes aside, or displaces, The size of some ships is expressed in terms of displacement, measured in tonnes of water.

Distillation

The process of heating a liquid until it gives off vapour, and then cooling the vapour until it condenses, or turns back to liquid. It is a common method of purifying liquids, or separating a mixture of liquids into separate fractions, as in the oil industry.

DNA

The organic compound deoxyribonucleic acid. It is one of the most important of all compounds because it is the substance inside living cells that controls the way they work and reproduce.

Electric circuit

A pathway along which electricity can flow.

Electricity

The science that deals with charged particles at rest and in motion. Electric charge is a fundamental property of matter. Atoms are made up of positively charged particles called protons and negatively charged particles called electrons. Electric charges can build up on certain objects such as rubber and plastics. Around such charges a field of force – an electric field – builds up. The study of charges like this and electric fields is often termed static electricity.

The flow of electrons – in a wire, say – creates the kind of electricity we are most familiar with: current electricity. A battery is a device that provides the 'pressure' that causes the electrons to flow through a conductor. We call the electron flow an electric current.

Electricity experiences a certain resistance when it flows through a conductor. The electrical resistance depends on the type of conductor. Copper and silver have very little resistance.

The units of electrical 'pressure', current and resistance are named after scientists who pioneered the study of electricity. For 'pressure' the unit is the volt (after Alessandro Volta); for current the unit is the amp (after André Marie Ampère); for resistance the unit is the ohm (after Georg Simon Ohm).

Ohm first stated a basic law of electricity – that in a circuit, voltage V equals current I times resistance R, or $V = IR$.

Hydroelectricity is electricity made by harnessing the energy of flowing water.

Electrode

The terminal through which current enters or leaves an electric cell or vacuum tube. By convention we say that current enters via the positive electrode, or anode, and that current leaves via the negative electrode, or cathode.

Electrolysis

Passing electricity through a substance in solution or when molten, when the substance is present as ions. It results in chemical changes taking place at the electrodes.

▼ **The principle underlying** electrolysis is used in electroplating. Copper goes into solution at the anode and is deposited on the key, the cathode.

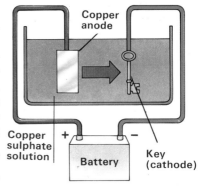

Copper anode

Copper sulphate solution

Battery

Key (cathode)

Electromagnetic spectrum
The range of wave radiations that consist of varying electric and magnetic fields. They differ in their wavelength, going from gamma rays (shortest wavelength) to radio waves (longest).

Electromagnetism
The science concerned with the relationship between electricity and magnetism. Electric motors and generators work on the principles of electromagnetism.

Electron
See Atom, Bonding, Electricity.

Element
A substance made up of atoms of the same atomic number – that is, with the same number of protons in the nucleus.

Energy
The capacity for doing work. Among the many forms of energy are chemical energy, electrical energy and heat energy. Potential energy is that possessed by a body because of its position; kinetic energy is the energy possessed by a moving body.

Evaporation
The change of state that occurs when a liquid changes into a vapour.

Fertilization
A stage in the reproduction of a living thing when a male cell joins with a female cell. From the fertilized cell an offspring will develop. In flowering plants, the male cells are contained in pollen, and the process is called pollination.

Fission
Splitting. Nuclear fission is the splitting of atoms, particularly uranium atoms. This process releases enormous energy as heat, light and radiation.

Force
Commonly, a push or a pull; something that can make a stationary body move, or change the speed or direction of a body when it is moving. Force is measured in newtons, named after Isaac Newton. One newton is the force that will produce an acceleration of 1 metre per second per second in a mass of 1 kilogram.

Friction
A force of resistance that acts between two moving bodies in contact and opposes motion.

Fusion
The change of state that occurs when a solid melts into a liquid. Nuclear fusion is a process in which light atoms (such as hydrogen) combine to make a heavier atom (such as helium). The process can take place only at temperatures of tens of millions of degrees. That is why it is often called thermonuclear fusion.

Genetics

The study of inheritance in living things – the passing on of characteristics from generation to generation – which takes place through the genes.

Gravity

The force of attraction between any bodies because of their mass. It is a fundamental property of matter and is the force that literally holds the universe together.

Greenhouse effect

A condition in which the Earth's atmosphere acts like a greenhouse and traps the Sun's heat. The effect is being brought about by the increasing concentration of heavy gases in the air, particularly carbon dioxide, which comes from burning fossil fuels. It is causing world temperatures to rise, which will in time change climates and may bring about the melting of the polar ice caps.

Heat

A form of energy, which is a measure of the kinetic energy possessed by the particles in a body. Latent heat is the heat that is taken in or given out when a substance changes its state from solid to liquid, liquid to gas and vice versa. See Evaporation, Fusion.

Hydraulics

A branch of science concerned with the behaviour of liquids and the way they transmit pressure.

Hydrocarbons

Compounds of hydrogen and carbon only. Petroleum, or crude oil, is made up almost entirely of hydrocarbons. So is natural gas, which is mainly the hydrocarbon methane.

Ion

A charged particle, consisting of an atom or a group of atoms that have acquired an electric charge by gaining or losing electrons. Positively charged ions are called cations and negatively charged ions are called anions.

Kinetic theory

The widely held theory which considers that matter is made up of particles (atoms or molecules). The kinetic energy of these particles increases with temperature. The kinetic theory can explain the changes in state from solid to liquid to gas, in terms of the increasing kinetic energy of the particles.

Latent heat

See Heat.

Light

The only form of electromagnetic radiation that our eyes can detect. White light is made up of different wavelengths, which we see as different colours when it is split up into a spectrum. All light rays travel at the same speed, some 300,000 kilometres per second. It is the fastest speed that can be reached.

Magnetism

A property possessed by iron and a few other metals, which makes them attract similar metals. Most ordinary magnets are made of iron, and they will attract pieces of iron and steel. The magnetism of a magnet appears to be concentrated at the ends, at the poles. One is called the south, and the other the north pole.

Magnetism is closely related to electricity, because a magnetic field is set up around a wire that is carrying electric current. See Electromagnetism.

Mass

The amount of matter in a body. It is measured in such units as grams and kilograms (thousands of grams). It is different from, but often confused with weight.

Metamorphosis

The change in form that insects go through in their life cycle; for example, egg, larva, pupa and adult.

Mineral

A natural chemical compound found in the ground, either by itself or in rocks. Every mineral has a fixed composition, no matter where it is found. It is often found in the form of crystals.

Modulation

The process of modifying a radio wave so that it can carry other signals. In radio, voice signals are combined with a radio carrier wave for transmission to radio receivers. In television, voice and picture signals are combined with a carrier wave for transmission.

Molecule

The smallest unit of a compound that can exist by itself and has the characteristic properties of that compound. It consists of two or more atoms chemically bonded together.

Neutron

An electrically neutral particle within the atom with a similar mass to a proton. See Atom.

Noble gas

One of the gases in Group 18 of the Periodic Table, which react with scarcely any other elements. They include helium, neon, krypton, argon and xenon. These elements are sometimes called the inert gases and rare gases.

Nucleus

The centre of an atom, which contains most of its mass. The centre of a living cell, which controls the way it operates.

Ore

A mineral that can profitably be extracted from the ground and processed into metal.

Organism

A living thing. A micro-organism is an organism that can be seen only in a microscope.

Oxidation

Originally, a process in which a substance combined with oxygen. The modern definition is that oxidation is a process in which a substance loses electrons. Oxidation is always accompanied by reduction, a process in which a substance gains electrons. The joint reduction-oxidation process is known as redox.

Particle

All matter is made up of particles. One basic particle is the atom. But atoms are not usually found by themselves. They are often joined together with other atoms to form larger particles called molecules. Sometimes atoms lose electrons when they combine and form ions. In some compounds, such as common salt, the basic particles are ions. Particles are also found within the atom itself, and are called sub-atomic particles.

pH

A measure of how acid or how alkaline a solution is. It is related to the concentration of hydrogen ions. The pH scale goes from 0 (strongly acidic) to 14 (strongly alkaline). Water is neutral at pH 7. See Acid, Alkali and the box on page 174.

Photosynthesis

The process by which plants make their food. They use the energy in sunlight to make carbon dioxide (taken in from the air) and water (taken in through the roots) combine to make food, in the form of sugar.

Plasma

Matter in a very high temperature state, in which it is present not as atoms, but as charged ions and electrons. Plasma is often called a fourth state of matter.

Polymerization

A chemical process in which many small molecules ('monomers') are made to combine together to form bigger, often very much bigger, molecules ('polymers').

Pressure

The force acting on a unit area, such as a square metre or square centimetre. Atmospheric pressure is the pressure exerted by the air above our heads. It works out at about 1 kilogram per square centimetre.

Proton

A particle with a positive electric charge, found in the nucleus of every atom. See Atom.

Radiation

Energy given off by a body, such as the Sun, in the form of electromagnetic rays or atomic particles. Alpha, beta and gamma radiation are given off by radioactive substances. See Radioactivity.

Radioactivity

A process in which unstable elements break down and give off radiation. There are three types of rays, or radiation – alpha, beta and gamma. Alpha rays consist of particles made up of two protons and two neutrons. Beta rays are electrons. Gamma rays are electromagnetic waves. The radiation comes from the nucleus of the atoms.

▲ **The international symbol** denoting 'Radioactive'.

Reflection

The rebound of a wave (such as sound and light) from a surface. Mirrors work because their silvered surfaces reflect light. When sound waves return to their source after reflection, they create an echo.

Refraction

The bending of light rays when they pass from one medium to another, such as from air to glass. Lenses work on the principle of refraction. They bend light rays and bring them to a focus, forming a sharp picture, or image. It may be a real image, which can be shown on a screen. Or it may be a virtual image, which can be viewed by the eye but cannot be seen on a screen.

Respiration

The process by which living things breathe. They take in oxygen in order to 'burn' the food they eat and produce energy. They give out carbon dioxide as a waste product.

Salt

The product (other than water) that results when an acid reacts with a base. For example, hydrochloric acid reacts with sodium hydroxide to form the salt sodium chloride (common salt). Chlorides, the salts of hydrochloric acid, are found widely in nature. So are sulphates, the salts of sulphuric acid, and nitrates, the salts of nitric acid.

Sciences

The word 'science' comes from the Latin word *scientia*, meaning knowledge. Scientists seek knowledge about the physical world we live in and about the living things that inhabit our world.

There are three main divisions of science – physics, the study of matter and energy; chemistry, the study of the properties and interactions of the substances that make up matter; and biology, the study of living things.

There are many other fields of scientific study within and in addition to these divisions. The main ones are shown in the accompanying table.

BRANCHES OF SCIENCE

Science	Study of

PHYSICAL SCIENCES

Science	Study of
Astronomy	The nature and evolution of the stars, planets, moons and all other heavenly bodies and space
Chemistry	The nature and properties of substances, and the way they react together
Electronics	The behaviour of electrons in a vacuum, in gases and in semiconductors
Engineering	The application of scientific principles to industry
Mechanics	The invention and construction of machines, their operation, and the calculation of their efficiency .
Metallurgy	The nature and properties of metals, their extraction, processing and shaping
Physics	The nature and properties of matter and energy

EARTH SCIENCES

Science	Study of
Geology	The Earth's composition, structure and surface, and the changes that take place there
Meteorology	The atmosphere and the weather
Mineralogy	Minerals, their location, mining and processing
Oceanography	Oceans, waves, currents, tides, trenches and ocean life
Palaeontology	Plant and animal fossils
Petrology	Formation, structure and composition of rocks

LIFE SCIENCES

Science	Study of
Anatomy	The form and structure of the body
Bacteriology	The growth and behaviour of bacteria
Biology	Living things
Botany	The nature and classification of plants
Ecology	The relationship between living things and the environment in which they live
Medicine	The causes, cure and prevention of diseases
Pharmacology	The effects, preparation and use of drugs
Physiology	The way living things work
Psychology	Mental processes and behaviour of humans and animals
Zoology	Animals, their behaviour and classification

MATHEMATICAL SCIENCES

Science	Study of
Logic	Reasoning by mathematics; used by computers
Mathematics	Numbers and quantities and their application; algebra, geometry, trigonometry, calculus
Statistics	Numerical information for the purposes of analysis

Semiconductor

A material that is able to conduct electricity to a limited degree. Conduction is brought about by the presence of certain impurities. Silicon is the best-known semiconductor. Its electrical properties depend on the presence of such impurities as phosphorus and boron.

Solution

A liquid containing one or more completely dissolved substances.

Sound

A wave motion that consists of physical vibrations – for example, of air molecules. When vibrating sound waves reach our ears, they cause our eardrums to vibrate, and we have the sensation of hearing.

We can hear sounds of only certain frequencies, or rates of vibration. Other animals can hear sounds of lower frequencies ('infrasounds') or of higher frequencies ('ultrasounds'). Bats, for example, navigate by means of ultrasounds.

At sea level, sound travels in air at a speed of about 1220 kilometres per hour. It travels slower the higher you go, as the air gets thinner.

In aircraft flight, the speed of sound ('sonic speed') is important. Aircraft travelling faster, at supersonic speed, need to be specially designed to combat the shock waves set up in the air at such speeds.

Space

The immense void through which the Earth, the planets and all the other heavenly bodies travel. Space begins above the atmosphere, at a height of 300 kilometres (about the height the space shuttle flies), where there is virtually no air left.

Space is almost, but not quite, a perfect vacuum. There are still a few gas particles and dust present. In some places, this interstellar matter forms denser nebulae ('clouds').

Spectrum

The spread of colour obtained when a beam of white light is passed through a prism or a diffraction grating. The colours – violet, indigo, blue, green, yellow, orange and red – represent light of slightly different wavelengths. They are spread out by a glass prism, say, because each colour is refracted (bent) through a slightly different angle as it passes into and out of the glass.

Surface tension

A force that exists at the surface of a liquid, which makes the liquid behave as if it has a skin. It is produced by the net downward pull on the surface particles of the liquid.

Thermodynamics

A branch of physics concerned with relationships between heat, work and other forms of energy.

Upthrust

The upward force a body experiences when it is floating in a fluid – a gas or a liquid.

Vacuum

A space that contains no matter. On Earth a perfect vacuum is impossible to achieve. And even space is not a perfect vacuum. See Space.

Valency

A measure of the combining, or bonding, power of an element. It is the number of hydrogen atoms (or their equivalent) which an atom of the element can combine with or displace in a chemical reaction.

Vapour

Below a certain point (the 'critical temperature'), a gas can be turned back into a liquid just by compressing it. In this state it is a vapour.

Viscosity

The property of a fluid that determines how it flows. In a highly viscous liquid, for example, the particles tend to stick together, and so the liquid flows slowly.

Voltage

See Electricity.

Volume

A measure of the space occupied by an object. It is expressed in such units as cubic centimetres or cubic metres.

Wave motion

A regular disturbance in a medium, or in space, that carries energy. Wave motions include water waves, sound waves and electromagnetic waves. Characteristics of a wave motion are its wavelength (the distance between the crest of one wave and the crest of the next), its frequency (the number of waves that pass by in a certain time), and its amplitude (the extent the wave is vibrating). For the same kind of wave, the product of the wavelength and frequency equals the velocity, or speed of the wave. See Electromagnetic spectrum.

Weight

The force a body experiences because of gravity. It is *not* the same as mass. The mass of a body – the amount of matter it contains – never changes. But the weight of a body changes when the pull of gravity changes. Confusingly, both mass and weight are usually expressed in the same units – grams and kilograms. See Gravity.

Work

Another term for energy, used in connection with moving forces. We say work is done when a force moves a certain distance. The amount of work done is measured in joules, named after James Prescott Joule. One joule is the work done when a force of 1 newton acts through a distance of 1 metre.

Index

Page numbers in *italic* type refer to illustrations. Page numbers in **bold** type refer to glossary entries.

ACKNOWLEDGEMENTS

The publishers wish to thank the following for supplying photographs for this book:

Page 6 ZEFA; 7 Biofotos; 8 Derby Museum & Art Gallery; 9 Science Photo Library; 11 ZEFA; 12 Solarfilma; 13 Grisewood & Dempsey; 14 Science Photo Library (top and middle) Ron Boardman (bottom); 15 Lu Jeffery; 16 ZEFA: 17 Pat Morris (top) Science Photo Library (bottom); 18 Science Photo Library; 20 Robin Kerrod; 21 Robin Kerrod; 23 LAT Photographic Ltd; 27 Science Photo Library; 30 Science Photo Library; 35 Greenpeace Communications Ltd; 37 Science Photo Library; 42 Science Photo Library; 43 ZEFA; 44 M. Holford; 45 Science Photo Library (top) J. Allan Cash (bottom); 46 Science Photo Library; 47 Science Photo Library; 48 John Walmsley; 49 Remote Source /L. Hughes; 51 Biofotos; 53 Science Photo Library; 56 Lu Jeffery; 57 Science Photo Library; 58 T & W Ide; 59 ZEFA; 62 Shell Photo Library; 65 NASA; 66 Transport & Road Research Authority; 67 Lu Jeffery; 69 J. Allan Cash; 70 Spectrum Colour Library; 71 Grisewood & Dempsey; 72 Allsport; 74 The Hutchinson Library; 77 NASA; 79 ZEFA; 82 Shell Photo Library; 84 Science Photo Library; 85 Allsport; 89 Science Photo Library; 92 ZEFA; 93 Science Photo Library; 94 ZEFA; 95 ZEFA; 96 J. Allan Cash; 97 ZEFA; 98 Grisewood & Dempsey; 100 Biofotos; 103 Perkin-Elmer Ltd; 105 ZEFA; 107 Science Photo Library (left) ZEFA (right); 109 ZEFA; 112 J. Allan Cash; 115 Science Photo Library; 116 ZEFA; 121 ARCO Ltd; 122 Michael Holford; 124 A.G.E. Fotostock; 128 ZEFA; 131 ZEFA; 132 Science Photo Library; 133 Courtesy, Elkington Ltd, Sheffield; 137 ZEFA; 139 Nevada Tourist Office; 141 Science Photo Library; 143 Ron Boardman; 145 Science Photo Library;153 Science Photo Library; 154 Marconi Radar Systems; 159 Science Photo Library; 162 ZEFA; 165 Science Photo Library; 166 NHPA; 167 NHPA; 171 ZEFA; 172 Camera Press; 173 Science Photo Library.

Illustrations by Robert and Rhoda Burns (pp. 13, 24–5, 26, 28–9, 30, 32, 33, 34–5, 38, 40–1, 47, 52, 54–5, 60–1, 63, 69, 73, 75, 83, 87, 90–1, 97, 98, 99, 101, 110, 113, 114, 119, 120–1, 123, 126–7, 131, 135, 137, 138–9, 140, 144–5, 148 (bottom), 152, 155, 161, 163, 173); Ralph Orme (pp. 31, 39, 133, 134).